INORGANIC SYNTHESES

Volume XVIII

Editor-in Chief

BODIE E. DOUGLAS

Department of Chemistry
University of Pittsburgh
Pittsburgh, Pennsylvania 15260

INORGANIC SYNTHESES

Volume XVIII

A Wiley-Interscience Publication
JOHN WILEY & SONS

New York Chichester Brisbane Toronto

Library of Congress Catalog Number: 39-23015

ISBN: 0-471-03393-6

Printed in the United States of America

10 9 8 7 6 5 4 3 2 1

PREFACE

Syntheses are reported for 113 compounds along with resolution procedures for some metal complexes. Several organic ligands are included, but their inclusion is warranted, since they are of interest primarily for the preparation of metal complexes. In several cases metal ions play an important role in the template synthesis of the ligands. There are several chapters of nonmetals, including an interesting phosphorus ylide and some of its metal complexes.

An important feature of *Inorganic Syntheses* is the independent checking of all procedures. We are grateful for the splendid cooperation of submitters and checkers and for the substantial time and effort required. It is evident that this is an international cooperative undertaking, since many of the submitters and checkers are from outside the United States. Contributions and suggestions are sought for future volumes. Directions for submitting syntheses follow this Preface. The editors of future volumes are listed opposite the title page.

Daryle Busch contributed a large group of syntheses and we were fortunate to have N. F. Curtis, another of the pioneers in the field of macrocycles, as a checker. I am grateful to the members of Inorganic Syntheses, Inc. who read the original manuscripts and made valuable suggestions. Major help with editing and proofreading was provided by W. Conard Fernelius, John C. Bailar, Jr., William L. Jolly, Warren H. Powell, and Thomas E. Sloan. Drs. Powell and Sloan of Chemical Abstracts Service have provided invaluable advice and help on nomenclature. Dr. Sloan prepared the Subject and Formula Indices. We hope that *Inorganic Syntheses* can help to establish good nomenclature practice so that inorganic chemists are able to benefit through better indices and improved access to the literature.

Bodie E. Douglas

CONTENTS

Chapter Two METAL CARBONYL COMPLEXES

Chapter Three OTHER COORDINATION COMPOUNDS

Chapter Four A PHOSPHORUS YLIDE AND SOME OF ITS METAL COMPLEXES

Chapter Five BORON AND ALUMINUM COMPOUNDS

Chapter Six GERMANIUM HYDRIDE DERIVATIVES

Chapter Seven PHOSPHORUS COMPOUNDS

Chapter Eight SULFUR COMPOUNDS

NOTICE TO CONTRIBUTORS

The *Inorganic Syntheses* series is published to provide all users of inorganic substances with detailed and foolproof procedures for the preparation of important and timely compounds. Thus the series is the concern of the entire scientific community. The Editorial Board hopes that all chemists will share in the responsibility of producing *Inorganic Syntheses* by offering their advice and assistance in both the formulation of and the laboratory evaluation of outstanding syntheses. Help of this kind will be invaluable in achieving excellence and pertinence to current scientific interests.

There is no rigid definition of what constitutes a suitable synthesis. The major criterion by which syntheses are judged is the potential value to the scientific community. An ideal synthesis is one that presents a new or revised experimental procedure applicable to a variety of related compounds, at least one of which is critically important in current research. However, syntheses of individual compounds that are of interest or importance are also acceptable. Syntheses of compounds that are available commercially at reasonable prices are not acceptable.

The Editorial Board lists the following criteria of content for submitted manuscripts. Style should conform with that of previous volumes of *Inorganic Syntheses*. The introductory section should include a concise and critical summary of the available procedures for synthesis of the product in question. It should also include an estimate of the time required for the synthesis, an indication of the importance and utility of the product, and an admonition if any potential hazards are associated with the procedure. The procedure should present detailed and unambiguous laboratory directions and be written so that it anticipates possible mistakes and misunderstandings on the part of the person who attempts to duplicate the procedure. Any unusual equipment or procedure should be described clearly. Line drawings should be included when they can be helpful. All safety measures should be stated clearly. Sources of unusual starting materials must be given, and, if possible, minimal standards of purity of reagents and solvents should be stated. The scale should be reasonable for normal laboratory operation, and any problems involved in scaling the procedure either up or down should be discussed. The criteria for judging the purity of the final product should be delineated clearly. The section of Properties should supply and discuss those physical and chemical characteristics that are relevant to judging the purity of the product and to permitting its handling and use in an intelligent manner. Under References, all pertinent literature citations should be listed in order. A style sheet is available from the Secretary of the Editorial Board.

The Editorial Board determines whether submitted syntheses meet the general specifications outlined above. Every synthesis must be reproduced satisfactorily in a laboratory other than the one from which it was submitted.

Each manuscript should be submitted in duplicate to the Secretary of the Editorial Board, Professor Jay H. Worrell, Department of Chemistry, University of South Florida, Tampa, FL 33620. The manuscript should be typewritten in English. Nomenclature should be consistent and should follow the recommendations presented in *Nomenclature of Inorganic Chemistry,* 2nd Ed., Butterworths & Co., London, 1970 and in *Pure Appl. Chem.,* **28**, No. 1 (1971). Abbreviations should conform to those used in publications of the American Chemical Society, particularly *Inorganic Chemistry.*

TOXIC SUBSTANCES

Recently it has become apparent that many common laboratory chemicals have subtle biological effects that were not suspected previously. These effects include latent carcinogenicity with induction periods of many years and teratogenicity to fetuses in addition to the previously recognized effects of chronic toxicity and of acute toxicity to certain susceptible individuals.

As a consequence of the new awareness of these hazards, the Occupational Safety and Health Administration of the United States government is in the process of establishing guidelines for the use of many chemicals. A sampling of the regulated chemicals is presented in the following table. Simple inspection of the list indicates that many widely used chemicals, such as benzene and carbon tetrachloride, should be handled with great care.

In light of the primitive state of our knowledge of the biological effects of chemicals, it is prudent that all the syntheses reported in this and other volumes of *Inorganic Syntheses* be conducted with rigorous care to avoid contact with all reactants, solvents, and products. The obvious hazards associated with these preparations have been delineated in each experimental procedure, but, at this point, it is impossible to foresee all possible sources of danger.

George W. Parshall

TOXIC SUBSTANCES 1976 OSHA Limits for Volatile Substances Often Found in the Inorganic Laboratory

Compound*	Maximum Allowable Exposure (ppm)	Compound	Maximum Allowable Exposure (ppm)
Acetone	1000	Fluorotrichloromethane	
Acetonitrile	40	(freon 11)	1000
Allyl chloride	1	Formaldehyde	3
Ammonia	50	Formic acid	5
Arsine	0.05	Heptane (*n*-heptane)	500
Benzene†	10	Hexane (*n*-hexane)	500
C Boron trifluoride	1	Hydrazine–skin†	1
Bromine	0.1	Hydrogen bromide	3
n-Butyl alcohol	100	C Hydrogen chloride	5
Carbon disulfide	20	Hydrogen cyanide	10
Carbon monoxide	50	Hydrogen fluoride	3
Carbon tetrachloride†	10	Hydrogen peroxide (90%)	1
Chlorine	1	Hydrogen selenide	0.05
Chlorine dioxide	0.1	C Hydrogen sulfide	20
Chlorobenzene		C Iodine	0.1
(monochlorobenzene)	75	Methyl Alcohol (Methanol)	200
C Chloroform†		Methylamine	10
(trichloromethane)	50	C Methyl bromide–skin	20
Cresol	5	Methyl chloride	100
Cyclohexane	300	Methylene chloride	75
Cyclopentadiene	75	Methyl iodide–skin†	5
Decaborane–skin	0.05	C Methyl mercaptan	10
Diborane	0.1	C Monomethyl hydrazine	
1,1-Dichloroethane	100	–skin	0.2
1,2-Dichloroethylene	200	Naphthalene	10
Diethylamine	25	Nickel carbonyl†	0.001
N,*N*-Dimethylacetamide–skin	10	Nitric acid	2
Dimethylamine	10	Nitric oxide (NO)	25
Dimethylformamide–skin	10	Nitrobenzene–Skin	1
Dimethyl sulfate–skin†	1	Nitrogen dioxide (NO_2)	1
Dioxane (diethylene		Nitrogen trifluoride	10
dioxide)–skin	100	Nitromethane	100
Ethyl acetate	400	Organo (alkyl)mercury	
Ethyl alcohol (ethanol)	1000	Oxalic acid	
Ethylamine	10	Oxygen difluoride	0.05
Ethyl bromide	200	Ozone	0.1
Ethyl ether	400	Pentaborane	0.005
Ethyl silicate	100	Pentane	1000
Ethylenediamine	10	Perchloroethylene	100
Ethylene dibromide		Perchloryl fluoride	3
(1,2-dibromoethane)†	20	Phenol–skin	5
Ethylene dichloride		Phenylhydrazine–skin	5
(1,2-dichloroethane)	5	Phosgene (Carbonyl	
Ethylene imine–skin†	0.5	chloride)	0.1
Ethylene oxide	50	Phosphine	0.3
Fluorine	0.1	Phosphoric acid	

Compound*	Maximum Allowable Exposure (ppm)	Compound	Maximum Allowable Exposure (ppm)
Phosphorus pentachloride		Sulfuryl fluoride	5
Phosphorus Pentasulfide		Tellurium hexafluoride	0.02
Phosphorus trichloride	0.5	Tetraethyl lead (as Pb)	
Pyridine	5	Tetrahydrofuran	200
Selenium compounds (As Se)		Tetramethyl lead (As Pb) –skin	
Selenium hexafluoride	0.05	Toluene	200
Stibine	0.1	Trichloroethylene	100
Sulfur dioxide	5	Triethylamine	25
Sulfur hexafluoride	1000	Vinyl chloride†	1
Sulfuric acid		Xylene (xylol)	100
Sulfur monochloride	1		
Sulfur pentafluoride	0.025		

*Entries preceded by a C are ceiling values that must not be exceeded at any time.
†Known or suspected to be a carcinogen.

INORGANIC SYNTHESES

Volume XVIII

Chapter One

MACROCYCLIC LIGANDS AND THEIR METAL COMPLEXES

Most of the known synthetic macrocyclic ligands have been prepared and characterized during the past decade. Most commonly they are quadridentates containing nitrogen donor atoms, although compounds containing oxygen and sulfur donor atoms are also known. The macrocyclic complexes of only a small number of metal ions (mostly first-row transition metal ions) have been studied in any detail, with greatest attention being given to rings of sizes ranging from 13 to 16 members. Several reviews[1-10] covering various aspects of the coordination chemistry of metal complexes of macrocyclic ligands have been published and reflect the increased attention being given to these compounds, especially those that could serve as models for more complex, biologically important macrocyclic systems. Many of the synthetic macrocycles are preferably prepared in the presence of metal ions (template reaction) to yield the metal complexes directly; for some macrocyclic ligands, these *in situ* syntheses remain their only mode of preparation. Alternately, the isolation of the free organic macrocycle prior to its use to prepare metal derivatives has been achieved in many cases.

1. 5,7,7,12,14,14-HEXAMETHYL-1,4,8,11-TETRAAZACYCLO-TETRADECA-4,11-DIENE (5,7,7,12,14,14-Me_6[14]-4,11-diene-1,4,8,11-N_4) COMPLEXES*

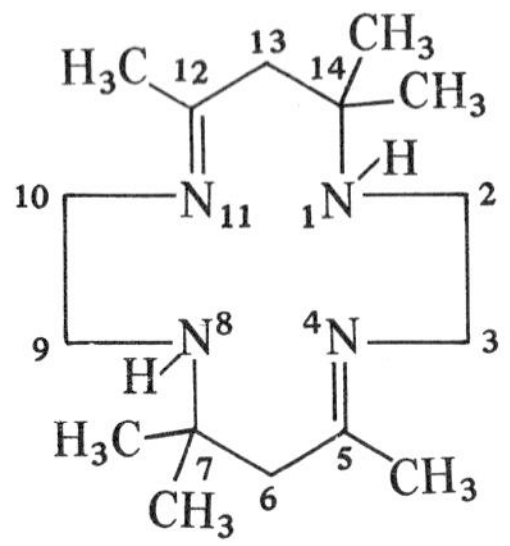

Submitted by A. MARTIN TAIT† and DARYLE H. BUSCH†
Checked by N. F. CURTIS‡

This ligand belongs to a group of widely studied synthetic macrocycles discovered by Curtis.[11] The first published procedure, involving the reaction of tris(ethylenediamine)nickel(II) perchlorate with acetone, is relatively arduous because (1) the condensation reaction occurs slowly over several days and (2) the major products[12-14] of the reaction are the chemically very stable positional isomers 5,7,7,12,14,14-hexamethyl-1,4,8,11-tetraazacyclotetradeca-4,11-diene (Me_6[14]-4,11-dieneN_4) and 5,7,7,12,12,14-hexamethyl-1,4,8,11-tetraazacyclotradeca-4,14-diene (Me_6[14]-4,14-dieneN_4). These isomers, which are similar in

*The nomenclature of macrocyclic ligands and the basis for abbreviations to represent the ligands are discussed in *J. Am. Chem. Soc.*, 94, 3397 (1972) and in *Inorg. Chem.*, **11**, 1979 (1972). Abbreviations for macrocyclic ligands are the subject of a current study by a joint working group of the IUPAC Commission on the Nomenclature of Inorganic Chemistry and the IUPAC Commission on the Nomenclature of Organic Chemistry. More complete abbreviations, giving all locant numbers, are given here. Shortened abbreviations omitting some locant numbers are used in equations and in the text where the meaning is clear. The unbracketed locant numbers in the *names* of all of the macrocyclic ligands referrring to substituents, unsaturation, and hetero atoms are those of the complete system, as defined by the rules for the nomenclature of organic ring systems. The abbreviations for the monocyclic ligands are derived directly from the systematic ring names and thus the locant numbers are the same as those in the names. However, the abbreviations for the polycyclic ligands are formed on the basis of a macromonocyclic ligand whose numbering is retained in the abbreviation for the polycyclic ligand for expressing the unsaturation, hetero atoms, and substituents (including attached rings) of the monocycle. Primed numbers, where needed, are used for ring components other than the monocycle. When different numbering is used for the name and for the abbreviation, both numerations are shown.

†Department of Chemistry, The Ohio State University, Columbus, OH 43210.
‡Chemistry Department, Victoria University of Wellington, Wellington, New Zealand.

appearance and physical properties, must be separated by fractional crystallization from the solvents, acetone, alcohol, water, or their mixtures.[12]

Curtis has discussed the preparation of complexes of this general class in a comprehensive review.[15] Condensation of several nickel and copper diamine complexes with a range of aliphatic aldehydes and ketones has led to the synthesis of a wide variety of macrocycles containing varying substituents and/or varying macrocyclic ring sizes.[16-19] The mechanism of the general reaction is not understood.

Complexes of $Me_6[14]$-4,11-dieneN_4 are, however, more conveniently prepared by reaction of the dianiono salt of the free organic ligand[20] with metal carbonates[21] or acetates[21,22] in methanol or water-methanol mixtures. The isolation of the free organic ligand salt is achieved by the reaction of the monoperchlorate, or similar salt of ethylenediamine, with acetone or mesityl oxide (4-methyl-3-penten-2-one).[20] The material can also be isolated from the reaction mixture produced by tris(ethylenediamine)iron(II) perchlorate and acetone.[21]

The preparations described here involve the reaction of the ciperchlorate or bis(trifluoromethanesulfonate) salt of $Me_6[14]$-4,11-dieneN_4 with the appropriate metal ion.

A. 5,7,7,12,14,14-HEXAMETHYL-1,4,8,11-TETRAAZACYCLOTETRADECA-4,11-DIENE BIS(TRIFLUOROMETHANESULFONATE) (5,7,7,12,14,14-$Me_6[14]$-4,11-diene-1,4,8,11-$N_4 \cdot 2CF_3SO_3H$).

$$2NH_2CH_2CH_2NH_2 \cdot CF_3SO_3H + 4CH_3COCH_3 \longrightarrow$$

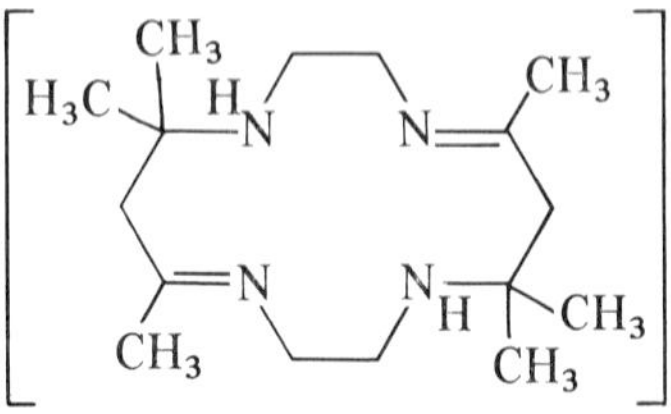

$$[\text{macrocycle}] \cdot 2CF_3SO_3H + 4H_2O$$

Procedure

To 300 mL of anhydrous methanol in a 1-L conical flask is added 20 g (0.33 mole) of ethylenediamine. Trifluoromethanesulfonic acid (50 g, 0.33 mole Aldrich Chemical Co.) is added very slowly from a dropping funnel to the rapidly stirred solution over a 1-hour period. The material is extremely hygroscopic, fumes in air, and reacts violently with most solvents. ■ **Caution.** *The use of a safety shield and protective clothing is recommended when handling this* *material.*) After the addition of the acid, the solution is filtered hot and cooled to room temperature. The methanol is removed on a rotary evaporator and the resultant cream-colored solid is dried *in vacuo* over P_4O_{10}. The anhydrous solid is dissolved in 300 mL of anhydrous acetone and the mixture is refluxed gently for 2 hours during which time the solution becomes deep red. (A small amount of white solid may be visible.) The solution is cooled to room temperature and reduced in volume to about 80 mL by rotary evaporation. Anhydrous diethyl ether (300 mL) is added, with stirring, to precipitate the white product, which is isolated by filtration. Addition of too much of the ether tends to produce an oil that will yield a solid upon addition of small amounts of tetrahydrofuran* and stirring. Additional material is obtained from the filtrate by alternate addition of ether and tetrahydrofuran (if an oil forms). The acetone, diethyl ether, and tetrahydrofuran mixture, upon reaching 500 mL in volume, is reduced to approximately 100 mL on a rotary evaporator. The dark-red solution is treated as previously indicated to obtain more of the product. This process is repeated several times. The combined product is stirred as a slurry in tetrahydrofuran to remove remnants of the oil, filtered, washed with tetrahydrofuran and then diethyl ether, and dried *in vacuo* over P_4O_{10}. The yield is 60-70 g (62-73%). *Anal.* Calcd. for $C_{16}H_{32}N_4 \cdot 2CF_3SO_3H$: C, 37.23; H, 5.86; N, 9.65; S, 11.07. Found: C, 37.20; H, 6.01; N, 9.69; S, 11.00.

B. 5,7,7,12,14,14-HEXAMETHYL-1,4,8,11-TETRAAZACYCLOTETRADECA-4,11-DIENE DIPERCHLORATE (5,7,7,12,14,14-Me_6[14]-4,11-diene-1,4,8,11-$N_4 \cdot 2HClO_4$).

$$2H_2NCH_2CH_2NH_2 \cdot HClO_4 + 4CH_3COCH_3 \longrightarrow 4H_2O +$$

$$\underline{NHCH_2CH_2N{=}C(CH_3)CH_2C(CH_3)_2NHCH_2CH_2N{=}C(CH_3)CH_2C}(CH_3)_2 \cdot 2HClO_4$$

Procedure

To 500 mL of anhydrous acetone in a 2-L beaker is added 20 g (0.33 mole) of ethylenediamine. The solution is stirred while 55.7 g (0.33 mole) of 60% perchloric acid is added slowly from a dropping funnel over a 30-minute period. (■ **Caution.** *The use of a safety shield is recommended. Perchlorate salts should all be regarded as potential hazards. Care should always be taken when handling either the dry solids or solutions in organic solvents. A very common source of problems is associated with evaporating solutions of perchlorates with*

*See *Inorg. Synth.*, **12**, 317 (1970), for precautions in handling tetrahydrofuran.

heat.) The solution becomes hot and an orange-red color develops. After addition of the acid, the solution is stirred rapidly and allowed to cool to room temperature. The fine white crystalline material is removed by filtration, washed thoroughly with acetone, and dried *in vacuo* over P_4O_{10}. Additional material can be isolated by stirring the reaction mixture for several days. The yield is 50-65 g, (63-78%). *Anal.* Calcd. for $C_{16}H_{32}N_4 \cdot 2HClO_4$: C, 39.92; H, 7.07; N, 11.64. Found: C, 40.06; H, 7.10; N, 11.63.

Properties

Both ligand salts are white crystalline materials that can be recrystallized from hot aqueous methanol. The perchlorate salt is insoluble in cold acetone, whereas the trifluoromethanesulfonate salt is considerably more soluble. The compounds show a strong broad band at 3150 cm^{-1} in their infrared spectra[21] due to the N—H vibration, and another weaker but broad band at 1544 cm^{-1} due to NH_2^+ vibration. The C=N stretching mode occurs as a strong sharp band at 1660 cm^{-1}. Other salts of the ligand with HI and HBF_4 have also been prepared.[23] The free base, $Me_6[14]$-4,11-dieneN_4, has also been obtained by displacing the ligand from nickel(II) with cyanide ion following the methods of Love and Powell[24] and Curtis.[25] The free base decomposes in the presence of moisture.

C. [*meso-* and *racemic*-(5,7,7,12,14,14-HEXAMETHYL-1,4,8,11-TETRAAZA-CYCLOTETRADECA-4,11-DIENE]NICKEL(II) PERCHLORATE

$$\mathrm{Ni(OOCCH_3)_2\cdot 4H_2O + Me_6[14]\text{-}4{,}11\text{-}dieneN_4\cdot 2HC10_4 \xrightarrow[\mathrm{CH_3OH}]{60^\circ}}$$

$$\mathrm{[Ni(\mathit{rac}\text{-}Me_6[14]\text{-}4{,}11\text{-}dieneN_4)](C10_4)_2 + 2CH_3COOH + 4H_2O}$$

$$\mathrm{[Ni(\mathit{rac}\text{-}Me_6[14]\text{-}4{,}11\text{-}dieneN_4)](C10_4)_2 \xrightarrow[\mathrm{H_2O}]{25^\circ}}$$

$$\mathrm{[Ni(\mathit{meso}\text{-}Me_6[14]\text{-}4{,}11\text{-}dieneN_4)](C10_4)_2}$$

*Procedure**

■ **Caution.** *See caution concerning perchlorates in Section 1-B.*

Nickel diacetate tetrahydrate (103 g,0.42 mole) is dissolved in 1.5L of methanol in a 3-L beaker and the ligand $Me_6[14]$-4,11-diene$N_4 \cdot 2HC10_4$ (180 g,

*The designations meso and racemic indicate whether the amine hydrogens are on the same or opposite sides, respectively, of the plane formed by the tour nitrogen atoms. This reflects the relative chiralities of the two amines.

0.37 mole) [Sec. 1-B] is added. A slight excess of nickel acetate is used to prevent contamination of the product with small amounts of unreacted ligand salt. Alternatively, equivalent amounts of Me_6[14]-4,11-dieneN_4·2CF_3SO_3H and $NaClO_4$ can be used. The solution is stirred for 1 hour at 60° and cooled rapidly in a refrigerator, and the yellow crystals of the racemic isomer are removed by filtration. The crystals are washed with methanol and diethyl ether and dried *in vacuo*. The yield is 170 g (84%). *Anal.* Calcd. for $C_{16}H_{32}N_4Cl_2O_8Ni$: C, 35.71; H, 5.95; N, 10.41. Found: (racemic isomer) C, 35.80 H, 6.01; N, 10.51. The isomerically pure meso isomer is prepared by keeping a suspension of the finely powdered racemic isomer at 25° under a saturated aqueous solution for 2 weeks with occasional agitation. The racemic isomer slowly dissolves and the meso isomer crystallizes.

Properties

Because of the restricted inversion of the asymmetric secondary amine functions, nickel(II) complexes of Me_6[14]-4,11-dieneN_4 can exist in both the meso and racemic forms,[12,14,26] which can be interconverted in solution. The diastereoisomers have been distinguished by optical rotation and NMR studies.[14] The NMR spectra show three equally intense resonances (at 2.67, 2.52 and 1.75 ppm for the racemic thiocyanate salt and at 2.69, 2.21, and 1.75 ppm for the meso thiocyanate salt in D_2O) assignable to the pairwise equivalent imine methyl groups (equatorial orientation) and to the axial and equatorial geminal methyl groups, respectively. The diastereoisomers each exhibit a single electronic absorption maximum between 21.0 and 23.0 kK[12,14,26] and infrared bands near 3200 and 1650 cm^{-1} assignable to the N—H and C=N functions, respectively. The complexes are all square planar, diamagnetic and behave as one-to-two electrolytes in water and in methanol. In neutral aqueous, acetone, or ethanolic solutions, the pure racemic and meso isomers convert to equilibrium mixtures. The ligand field strength of the macrocyclic ligand has been calculated[27] to be 1569 cm^{-1}, making the ligand one of the most strongly coordinating of the synthetic macrocyclic quadridentates on nickel(II). The NCS^-, PF_6^-, and BF_4^- salts have also been prepared and characterized.

D. BIS(ACETONITRILE)(5,7,7,12,14,14-HEXAMETHYL-1,4,8,11-TETRAAZACYCLOTETRADECA-4,11-DIENE)IRON(II) TRIFLUOROMETHANESULFONATE

$$[Fe(CH_3CN)_6](CF_3SO_3)_2 + Me_6[14]\text{-4,11-diene}N_4\cdot 2CF_3SO_3H + 2(C_2H_5)_3N \longrightarrow$$

$$2[(C_2H_5)_3NH](CF_3SO_3) + [Fe(Me_6[14]\text{-4,11-diene}N_4)(CH_3CN)_2](CF_3SO_3)_2$$

*Procedure**†

Thirty grams (0.2 mole) of trifluoromethanesulfonic acid‡ from a freshly opened bottle is added slowly from a dropping funnel to 200 mL of dry degassed acetonitrile§ in a 500-mL, round-bottomed flask equipped with a ground-glass joint and further deoxygenated by the passage of dry nitrogen. The flask is then tightly sealed and taken into a nitrogen atmosphere glove box and transferred to a 1-L Erlenmeyer flask equipped with a ground-glass joint.‖ Six grams of finely divided iron powder (slightly in excess of 0.1 mole) is added and the iron powder is agitated by activating the stirring mechanism of a hot/stir plate. The solution is warmed gently for a few minutes until reflux and is then kept in a state of gentle reflux for 24 hours. An air condenser should be used to minimize evaporation, and the volume of the solution should be maintained at 200 mL by addition of acetonitrile as required. The pale-yellow solution of $[Fe(CH_3CN)_6]$-$(CF_3SO_3)_2$ so obtained is filtered into a 1-L flask through filter paper¶ to remove the excess iron powder and is again diluted to 200 mL by addition of acetonitrile. Fifty-eight grams (0.1 mole) of the very dry ligand salt $Me_6[14]$-4,11-dieneN_4·$2CF_3SO_3H$ [Sec. 1-A] is added with stirring, followed by the addition of 50 mL of dry, degassed triethylamine. The deep red-purple solution is reduced to a volume of approximately 100 mL by stirring under vacuum and the pale-pink material that precipitates is removed by filtration, washed with a 50:50 mixture

*The checker used one-third the scale described here.

†This complex cation was first prepared by the reaction of anhydrous iron(II) perchlorate with the perchlorate salt of the macrocyclic ligand.[28] However, the conditions under which this product is obtained are extremely hazardous.[28] Related experiments under similar conditions have produced violent explosions.[29] The perchlorate compound, as well as many other iron derivatives containing the perchlorate anion, are both thermal and shock sensitive and detonate with sharp reports.

‡Trifluoromethanesulfonic acid once opened to the atmosphere slowly turns brown or black, and such solutions should not be used in the preparation of this iron complex for which strictly anhydrous conditions are required. The acid should be used as soon as possible after air exposure. It can, however, be stored under nitrogen in an all-glass container for short periods. The acid destroys Bakelite caps and these should never be used in storage.

§Acetonitrile was dried over molecular sieves and then distilled from CaH_2 under nitrogen after refluxing for 1 hour under nitrogen. Diethyl ether was refluxed over CaH_2 under nitrogen and then distilled. Triethylamine was dried over KOH and distilled from KOH under nitrogen.

‖The primary difficulty in preparing iron(II) complexes of aliphatic amine ligands is related to the very great tendency of iron to form hydroxo species and their pronounced tendency to oxidize to various iron(III) oxo species in the presence of traces of water. To circumvent these difficulties the synthetic procedure for this iron(II) complex should be carried out under rigorously anhydrous conditions in an inert atmosphere.

¶If the solution of iron powder and trifluoromethanesulfonic acid in acetonitrile is to be filtered through a sintered-glass filter, the addition of a small amount of Filter Aid to the reaction mixture will prevent the sinter from becoming clogged with the unreacted iron powder.

of dry acetonitrile–diethyl ether, and dried by suction. The product is removed from the glove box and further dried *in vacuo* over P_4O_{10}. The yield is 50 g* *Anal.* Calcd. for $C_{22}H_{38}N_6S_2O_6F_6Fe$: C, 36.88; H, 5.31; N, 11.73; S, 8.94. Found: C, 36.91; H, 5.27; N, 11.65; S, 8.90.

Properties

The acetonitrile complex is low-spin and six-coordinate (C≡N infrared band of CH_3CN occurs at 2275 cm^{-1}) and behaves as a one-to-two electrolyte in nitromethane and in acetonitrile.[28] The solid compound decomposes slowly in the presence of oxygen and oxidizes rapidly in solution upon exposure to air. Such solutions undergo a complex series of dehydrogenation reactions to yield new iron(II) complexes containing macrocyclic ligands with greater degrees of unsaturation.[32,33] The visible spectrum of this acetonitrile complex shows one *d-d* transition at 19.6 kK, with more intense absorptions appearing at higher energies. The NMR spectrum shows three equally intense methyl resonances at 2.42, 1.39, and 1.00 ppm assignable to the equatorial imine methyl group and to the equatorial and axial geminal methyl groups, respectively. The resonance of the methyl group of acetonitrile occurs at 1.95 ppm.

The infrared spectrum of the complex shows a very intense N–H stretching absorption at about 3220 cm^{-1} and an intense imine C=N stretching absorption in the 1670-1640 cm^{-1} region.

Metathetical reactions on the complex $[Fe(Me_6[14]\text{-}4,11\text{-}dieneN_4)(CH_3CN)_2]^{2+}$ lead to a series of low-spin six-coordinate (with coordinated NCS^-) and high-spin five-coordinate (Br^-, I^-, OH^-) complexes. Low-spin iron(III) complexes are also known.[28]

References

1. D. H. Busch, *Record. Chem. Progr.*, **25**, 107 (1964).
2. D. St. C. Black and E. Markham, *Rev. Pure Appl. Chem.*, **15**, 109 (1965).
3. A. B. P. Lever, *Advan. Inorg. Chem. Radiochem.*, 7, 27 (1965).
4. D. H. Busch, *Helv. Chim. Acta* (fasciculus extraordinarius Alfred Werner), **1967**, 174.

*Although the use of a glove box is recommended for the preparation of this complex, it can be prepared on the bench top under a blanket of nitrogen. Various types of glass apparatus[30,31] have been designed for the purpose of filtering in an inert atmosphere, but such apparatus is not as convenient to use as a glove box. Yields are generally lower and the product less pure.

5. N. F. Curtis, *Coord. Chem. Rev.*, **3**, 3 (1968).
6. L. F. Lindoy and D. H. Busch, *Preparative Inorganic Reactions,* W. L. Jolly, ed., Vol. 6, John Wiley and Sons, Inc., New York, 1971, p. 1.
7. D. H. Busch, K. Farmery, V. Goedken, V. Katovic, A. C. Melnyk, C. R. Sperati, and N. Tokel, *Adv. Chem. Ser.*, **100**, 44 (1971).
8. J. J. Christensen, D. J. Eatough, and R. M. Izatt, *Chem. Rev.*, **74**, 351 (1974).
9. L. F. Lindoy, *Chem. Soc. Rev.*, **4**, 421 (1975).
10. D. H. Busch, D. G. Pillsbury, F. V. Lovecchio, A. M. Tait, Y. Hung, S. Jackels, M. C. Rakowski, W. P. Schammel, and L. Y. Martin, American Chemical Society Symposium Series, **38**, 32 (1977).
11. N. F. Curtis, *J. Chem. Soc.*, **1960**, 4409; N. F. Curtis and D. A. House, *Chem. Ind. (London)*, **42**, 1708 (1961).
12. N. F. Curtis, Y. M. Curtis, and H. J. K. Powell, *J. Chem. Soc. (A)*, **1966**, 1015.
13. R. R. Ryan, B. T. Kilbourn, and J. D. Dunitz, *Chem. Commun.*, **1966**, 910; M. F. Bailey and I. E. Maxwell, *Chem. Commun.*, **1966** 908, B. T. Kilbourn, R. R. Ryan and J. D. Dunitz, *J. Chem. Soc. (A)*, **1969**, 2407.
14. L. G. Warner, N. J. Rose, and D. H. Busch, *J. Am. Chem. Soc.*, **90**, 6938 (1968); **89**, 703 (1967).
15. N. F. Curtis, *Coord. Chem. Rev.*, **3**, 3 (1968).
16. M. M. Blight and N. F. Curtis, *J. Chem. Soc.*, **1962** 1402, 3016; N. F. Curtis, *J. Chem. Soc., Dalton,* **1972**, 1357; **1973**, 863; **1974**, 347; D. F. Cook and N. F. Curtis, *J. Chem. Soc., Dalton,* **1973**, 1076.
17. D. A. House and N. F. Curtis, *J. Am. Chem. Soc.*, **84**, 3248 (1962); **86**, 223, 1331 (1964).
18. R. W. Hay and G. A. Lawrence, *J. Chem. Soc. Dalton,* **1975**, 1466.
19. R. A. Kolinski and B. Korybut-Daszkiewicz, *Bull. Acad. Polon. Sci.*, **17**, 13 (1969); *Inorg. Chim. Acta,* **14**, 237 (1975).
20. N. F. Curtis and R. W. Hay, *Chem. Commun.*, **1966**, 524.
21. N. Sadasivan and J. F. Endicott, *J. Am. Chem. Soc.*, 88, 5468 (1966).
22. L. G. Warner, N. J. Rose, and D. H. Busch, Abstracts of the 52nd National Meeting of the American Chemical Society, Paper No. 142, New York, 1966.
23. L. G. Warner, Ph.D. Thesis, The Ohio State University, 1968.
24. J. L. Love and H. K. J. Powell, *Inorg. Nucl. Chem. Lett.*, **3**, 113 (1967).
25. N. F. Curtis, *J. Chem. Soc.*, **1964**, 2644.
26. N. F. Curtis and Y. M. Curtis, *J. Chem. Soc. (A)*, **1966**, 1653.
27. C. R. Sperati, Ph.D. Thesis, The Ohio State University, 1971.
28. V. L. Goedken, P. H. Merrell, and D. H. Busch, *J. Am. Chem. Soc.*, **94**, 3397 (1972).
29. R. C. Dickinson and G. R. Long, *Chem. Eng. News,* **July, 1970,** 6.
30. W. L. Jolly, *Inorg. Synth.*, **11**, 116 (1968).
31. J. J. Eisch and R. B. King (eds.), *Organometallic Chemistry,* Vol. 1, Academic Press, Inc., New York, 1965, p. 55.
32. A. M. Tait and D. H. Busch, *Inorg. Chem.*, **15**, 197 (1976).
33. V. L. Goedken and D. H. Busch, *J. Am. Chem. Soc.*, **94**, 7355 (1972).

2. 5,5,7,12,12,14-HEXAMETHYL-1,4,8,11-TETRAAZACYCLOTETRADECANE (5,5,7,12,12,14-Me_6[14]ane-1,4,8,11-N_4) COMPLEXES

Submitted by A. MARTIN TAIT* and DARYLE H BUSCH*
Checked by N. F. CURTIS†

This fully saturated macrocyclic ligand was prepared initially by the reduction of the precursor nickel complex [Ni(5,7,7,12,14,14-Me_6[14]-4,11-diene-1,4,8,11-N_4)]$(ClO_4)_2$ [Sec. 1-C] with sodium tetrahydroborate(1−)[1,2] or with nickel/aluminum alloy.[2] Theoretically, the nickel(II) complex of Me_6[14]aneN_4 can exist in 20 isomeric forms, depending on the configurations of the two asymmetric carbon centers (meso or racemic) and the four asymmetric nitrogen centers.[2,3] Only three isomers have been characterized. The fully saturated ligand can be removed from nickel in the presence of cyanide ion,[1] allowing for the preparation of other transition-metal complexes by direct combination of the appropriate metal salt with the free ligand. However, the more direct syntheses of the tetraaza macrocycle Me_6[14]aneN_4 without the interaction of template or other ligand reactions, is easier and more convenient for the preparation of all these complexes.[4]

A. *meso-* and *racemic-*(5,5,7,12,12,14-HEXAMETHYL-1,4,8,11-TETRAAZACYCLOTETRADECANE) HYDRATE (5,5,7,12,12,14-Me_6[14]ane-1,4,8,11-$N_4 \cdot xH_2O$)

$$\overline{NHCH_2CH_2N{=}C(CH_3)CH_2C(CH_3)_2NHCH_2CH_2N{=}C(CH_3)CH_2C}(CH_3)_2 \cdot 2HClO_4$$

$$\downarrow NaBH_4$$

$$\underline{NHCH_2CH_2NHC(CH_3)_2CH_2CH(CH_3)NHCH_2CH_2NHC(CH_3)_2CH_2C}H(CH_3) \cdot xH_2O$$

*Department of Chemistry, The Ohio State University, Columbus, OH 43210.
†Chemistry Department, Victoria University of Wellington, Wellington, New Zealand.

x = 2 for meso isomer
x = 1 for racemic isomer

Procedure

■ **Caution.** *See caution concerning perchlorates in Section 1-B.*

To 500 mL of methanol in a 2-L beaker is added 100 g (0,21 mole) of 5,7,7,12,14,14-Me_6[14]-4,11-diene-1,4,8,11-$N_4 \cdot 2HClO_4$ [Sec. 1-B]. The solution is stirred and 19 g (0.63 mole) of sodium tetrahydroborate(1−) and 16.5 g (0.42 mole) of sodium hydroxide are added in alternate small portions over a 1-hour period. The addition should be conducted in a well-ventilated fume hood (hydrogen is evolved) and in a cold-water bath to moderate the temperature, if necessary. After the addition is complete, the solution is stirred for 1 hour at room temperature and then heated to reflux for 15 minutes. After it is cooled, 50 g of sodium hydroxide in 1 L of water is added and the solution is stirred until precipitation of the product is complete (usually about 1 hr) and then filtered. The white product is washed with cold water and air-dried overnight. Additional material can be recovered from the filtrate by volume reduction. The yield is 52 g (88%). The product is a 50:50 mixture of the meso and racemic isomers that can be separated by fractional crystallization from methanol as detailed below.*

Meso Isomer

The product (52 g) from the previously described reaction is dissolved in 600 mL of methanol at reflux temperature and filtered hot to remove some intractible brown material. The filtrate is diluted to 600 mL with methanol and the solution is reheated to reflux. Water (400 mL) is added to the hot solution, which is then stirred and cooled to room temperature. The white finely powdered material that precipitates is the pure meso isomer. It is removed from the mixture by filtration, washed with cold water, and dried *in vacuo* over P_4O_{10}. The yield is 20-23 g (30-34%). *Anal.* Calcd. for $C_{16}H_{36}N_4 \cdot 2H_2O$: C, 60.00; H, 12.50; N, 17.50. Found: C, 60.11; H, 12.58; N, 17.27.

Racemic Isomer

To the filtrate remaining after removal of the meso isomer is added a further 200 mL of water. The solution is stirred rapidly for about 30 minutes during which time a precipitate forms. This material (about 8 g or less) is a mixture of both meso and racemic isomers and is removed by filtration. The filtrate that remains

*The designations meso and racemic refer to the asymmetric carbon atoms at positions 7 and 14. Chiralities for these carbon atoms in the meso ligand are *R* and *S* while for the racemic isomer they are *R,R* or *S,S*.

is evaporated to near dryness on a rotary evaporator and the white product is isolated. This is the pure racemic isomer and it is washed with a small amount of cold water and dried *in vacuo* over P_4O_{10}. The yield is about 21 g (33%). Yields of the racemic isomer can be improved by extraction of the final filtrate and washings with diethyl ether, followed by evaporation of the diethyl ether extract. *Anal.* Calcd. for $C_{16}H_{36}N_4 \cdot H_2O$: C, 63.58; H, 12.58; N, 18.54. Found: C, 63.81; H, 12.58; N, 18.80.

Properties

The isomeric meso and racemic macrocycles crystallize initially as the dihydrate and monohydrate, respectively, melting points of 146-148 and 97-105°. The compounds can be dehydrated by prolonged exposure to P_4O_{10} *in vacuo* but revert to their hydrates on exposure to the atmosphere. The meso isomer is insoluble in cold water, sparingly soluble in ethers, but readily soluble in alcohols. The racemic isomer is appreciably more soluble in all these solvents and is conveniently purified by extraction from alkaline aqueous solutions into diethyl ether, from which it can be recrystallized. The isomers can be distinguished readily by their infrared spectra.[1,4] The meso isomer shows four bands at 1248, 1273, 1284, and 1305 cm^{-1} in the 1240-1310 cm^{-1} region, while the racemic isomer shows only three bands at 1259, 1275, and 1305 cm^{-1}. The NH absorptions occur near 3300 cm^{-1} while the broad bands near 3400 and 1700 cm^{-1} are associated with the hydrogen-bonded water molecules. The mass spectra both show the parent ion peak at m/e 284, with a peak at m/e 269 corresponding to the loss of one methyl group. The NMR spectrum of the meso ligand shows a singlet at 1.07 ppm with a shoulder at 1.06 ppm assignable to the geminal methyl groups and a doublet centered at 0.98 ppm due to the equivalent lone methyl groups. The spectrum of the racemic ligand is almost identical with comparable bands at 1.10, 1.08, and 0.98 ppm (coupling constant = 5.8 cps) respectively.

B. [*meso*-(5,5,7,12,12,14-HEXAMETHYL-1,4,8,11-TETRAAZACYCLOTETRADECANE]NICKEL(II) PERCHLORATE

$$Ni(OOCCH_3)_2 \cdot 4H_2O + meso\text{-}(Me_6[14]aneN_4) \cdot 2H_2O + 2NaClO_4 \xrightarrow{CH_3OH}$$
$$[Ni(meso\text{-}Me_6[14]aneN_4)](ClO_4)_2 + 2NaOOCCH_3 + 4H_2O$$

Procedure

■ **Caution.** *See caution concerning perchlorates in Section 1-B.*

To a solution of 6.8 g (0.03 mole) of nickel diacetate tetrahydrate in 125 mL of methanol is added 8.0 g (0.025 mole) of *meso*-Me_6[14]aneN$_4$·$2H_2O$ (Sec. 2-A). The solution is stirred and heated at 60° for 1 hour. The red-brown solution is cooled to room temperature and filtered. The filtrate is warmed to redissolve the precipitated purple diacetato complex and 6.1 g (0.05 mole) of sodium perchlorate is added to produce an immediate precipitate of the yellow meso complex.* The solution is stirred for 30 minutes and filtered. The solid is washed with methanol and then diethyl ether and is dried *in vacuo* over P_4O_{10}. Yield is 11 g (81%). *Anal.* Caled. for $C_{16}H_{36}N_4Cl_2O_8Ni$: C, 35.45; H. 6;69; N, 10.34. Found: C, 35.61; H, 6.49; N, 10.40.

Properties

Based on symmetry and steric arguments, the meso ligand is most likely to coordinate only with its four nitrogen atoms arranged in a single plane, while the racemic ligand may coordinate in either this fashion or in a folded form.

The α- and β-racemic isomers are diastereoisomers related through configurational differences among the coordinated asymmetric nitrogens.[2,3] The α-isomer in the square planar form has one trans pair of secondary amine hydrogens directed above, with the other trans pair of secondary amine hydrogens below the plane of the nitrogen atoms. The β-isomer in the planar form has all four secondary amine hydrogens on the same side of the nitrogen plane.

All the complexes show a strong sharp band near 3200 cm^{-1} due to the N–H vibrations but no imine (C=N) absorptions. The perchlorate complexes are all diamagnetic and square planar in the solid and in solution, showing one absorption in the visible spectrum near 22.00 kK.

The NMR of the complex [Ni(*meso*-Me_6[14]aneN$_4$)](ClO_4)$_2$ shows a doublet at 1.08, 1.15 ppm due to the lone equatorial methyl groups and singlets at 1.15 and 1.72 ppm assignable to the geminal equatorial and axial methyl groups, respectively. The NMR spectrum of β-[Ni(*racemic*-Me_6[14]aneN$_4$)](ClO_4)$_2$ shows the analogous resonances at 0.97, 1.07, 1.11, and 2.12 ppm, respectively. Detailed conformational analyses of the complexes have been performed.[2,3] Analysis[5] of the visible spectrum[5-7] of [Ni(*meso*-Me_6[14]aneN$_4$)(NCS)$_2$] indicates that the ligand Me_6[14]aneN$_4$ has a ligand field strength Dq^{xy} of 1398 cm^{-1}.

* Addition of *racemic*-Me_6[14]aneN$_4$·H_2O to nickel diacetate in methanol produces the blue-violet six-coordinate acetato complex [Ni(*racemic*-Me_6[14]aneN$_4$)(CH_3COO)]ClO_4 in which the macrocycle is coordinated in a folded form. Addition of perchloric acid to this acetato complex in water produces the yellow square planar complex α-[Ni(*racemic*-Me_6[14]aneN$_4$)](ClO_4)$_2$.[2,3] which isomerizes in solution to the β-racemic form, by way of inversion of two nitrogen atoms.

C. DIBROMO[*meso*-(5,5,7,12,12,14-HEXAMETHYL-1,4,8,11-TETRAAZA-CYCLOTETRADECANE)]COBALT(III) PERCHLORATE

$$CoBr_2 \cdot 6H_2O + meso\text{-}Me_6[14]aneN_4 \cdot 2H_2O + NaClO_4 \xrightarrow[O_2]{H^+}$$

$$trans\text{-}[Co(meso\text{-}Me_6[14]aneN_4)Br_2]ClO_4 + 6H_2O$$

Procedure

■ **Caution.** *See caution concerning perchlorates in Section 1-B.*

meso-$Me_6[14]aneN_4 \cdot 2H_2O$ (9.6 g, 0.03 mole) is dissolved in 250 mL of methanol. Glacial acetic acid (1.8 mL, 0.03 mole) is added, followed by 9.8 g (0.03 mole) of cobalt dibromide hexahydrate. The wine-red solution is aerated for several hours to produce a green color. The solution is evaporated to dryness on a rotary evaporator and the residue is dissolved in 5% aqueous hydrobromic acid. Excess sodium perchlorate (10 g) is added with stirring and the pale-green precipitate is removed by filtration and washed with ethanol. The solids are dissolved in hot dilute ammonia (the dihydroxo compound is formed in solution) and the solution is filtered hot. The filtrate is acidified with hydrobromic acid, and 2 g of sodium perchlorate is added. The solution is digested on a steam bath for 30 minutes to insure reconversion to the dibromo complex. The product is isolated by cooling and filtration and is washed with ethanol before drying *in vacuo* over P_4O_{10}. The yield is 14 g (78%). *Anal.* Calcd. for $C_{16}H_{36}N_4Br_2ClO_4Co$: C, 31.88; H, 5.98; N, 9.30; Br, 26.54. Found: C, 31.65; H, 5.93; N, 9.39; Br, 26.19.

Properties

The cobalt(III) complex $[Co(meso\text{-}Me_6[14]aneN_4)Br_2]ClO_4$ is diamagnetic and six-coordinate.[8] It is slightly soluble in alcohols and water but soluble in acetonitrile and nitromethane, in which it behaves as a one-to-one electrolyte. The visible spectrum of the dibromo complex shows one *d-d* transition at 14.6 kK. Analysis of the spectra of several of the related complexes after the method of Wentworth and Piper[9] leads to a ligand field strength Dq^{xy} of 2460 cm^{-1}, suggesting that the macrocyclic ligand is one of the weakest 14-membered macrocycles studied for cobalt(III).

All the cobalt complexes of this ligand show bands near 3200 cm^{-1} in their infrared spectra assignable to the N—H modes.[8,10] A wide variety of cobalt(III) complexes[8,11,12] have been prepared and characterized with this saturated ligand with the analogous racemic ligand.[8,11]

References

1. N. F. Curtis, *J. Chem. Soc. (A)*, **1965**, 924; **1967**, 2644.
2. L. G. Warner and D. H. Busch, *J. Am. Chem. Soc.*, **91**, 4092 (1969).
3. L. G. Warner and D. H. Busch, *Coordination Chemistry*, Papers presented in honor of Professor John C. Bailar, Jr., Plenum Press, New York, 1969.
4. A. M. Tait and D. H. Busch, *Inorg. Nucl. Chem. Lett.*, 8, 491 (1972).
5. C. R. Sperati, Ph.D. Thesis, The Ohio State University, 1971; L. Y. Martin, C. R. Sperati, and D. H. Busch, *J. Am. Chem. Soc.*, **99**, 2968 (1977).
6. A. B. P. Lever, *Coord. Chem. Rev.*, **3**, 119 (1968).
7. D. A. Rowley and R. S. Drago, *Inorg. Chem.*, **1**, 1795 (1968).
8. P. O. Whimp and N. F. Curtis, *J. Chem. Soc. (A)*, **1966**, 867, 1827.
9. R. A. D. Wentworth and T. S. Piper, *Inorg. Chem.*, **4**, 709 (1965).
10. L. G. Warner, Ph.D. Thesis, The Ohio State University, 1968.
11. A. M. Tait and D. H. Busch, unpublished results.
12. N. Sadasivan, J. A. Kernohan, and J. F. Endicott, *Inorg. Chem.*, **6**, 770 (1967).

D. BIS(ACETONITRILE)[*meso*-(5,5,7,12,12,14-HEXAMETHYL-1,4,8,11-TETRAAZACYCLOTETRADECANE)IRON(II) TRIFLUOROMETHANE-SULFONATE

Submitted by A. MARTIN TAIT*, DENNIS P. RILEY*, and DARYLE H. BUSCH*
Checked by N. F. CURTIS†

$$Fe(OOCCH_3)_2 + meso\text{-}Me_6[14]aneN_4 + 2CF_3SO_3H \xrightarrow{CH_3CN}$$
$$[Fe(meso\text{-}Me_6[14]aneN_4)(CH_3CN)_2](CF_3SO_3)_2 + 2CH_3COOH$$

Procedure‡

To 800 mL of dry degassed acetonitrile§ warmed to 50-60° is added 5.78 g

*Department of Chemistry, The Ohio State University, Columbus, OH 43210.
†Chemistry Department, Victoria University of Wellington, Wellington, New Zealand.

‡(1) The analogous tetrafluoroborate salt is prepared in a similar manner,[3] using 30% tetrafluoroboric acid [hydrogen tetrafluoroborate(1−)] in place of trifluoromethanesulfonic acid. This deep-purple complex readily loses the coordinated acetonitrile groups on drying *in vacuo* to produce a pale-blue air-sensitive material formulated as $[Fe(meso\text{-}Me_6[14]aneN_4)(BF_4)_2]$ containing coordinated tetrafluoroborate groups.[1] This material reverts to the purple acetonitrile adduct upon exposure to acetonitrile. The use of perchloric acid to synthesize the perchlorate salt should be avoided because of the hazardous nature of this material.[3,4]

§Acetonitrile was dried over molecular sieves and then distilled from CaH_2 under nitrogen after reflux for 1 hour under nitrogen. Tetrahydrofuran was dried over KOH and then refluxed over and distilled from CaH_2 under nitrogen.

(0.033 mole) of anhydrous iron(II) acetate under nitrogen. The resulting suspension is stirred for 10 minutes prior to the addition of 9.45 g (0.033 mole) of very dry *meso*-$Me_6[14]aneN_4$ (Sec. 2-A). The solids dissolve slowly to form first a light-green solution, which after about 30 minutes turns to a blue-green with a small amount of green suspended solid present. To this suspension is added slowly 10.0 g (0.067 mole) of trifluoromethanesulfonic acid.* The green diacetato salt dissolves immediately and the deep-purple bis(acetonitrile) adduct forms. The solution is filtered hot under nitrogen, using suitably designed glass filtration apparatus,[1,2]† and the solution volume is reduced by stirring *in vacuo* to about 150 mL. The deep-purple crystalline solid is removed by filtration under a stream of nitrogen and washed with a small amount of cold degassed acetonitrile. More product can be recovered from the filtrate by addition of about 200 mL of degassed dry tetrahydrofuran, followed by about 20 mL of diethyl ether. The yield is 19-21 g (79-90%). For analytical purposes, a small amount of the material can be recrystallized from hot degassed acetonitrile under nitrogen upon volume reduction. *Anal.* Calcd. for $C_{22}H_{42}N_6S_2O_6F_6Fe$: C, 36.67; H, 5.88; N, 11.67; S, 8.89. Found: C, 36.49; H, 5.89; N, 11.46; S, 8.92.

Properties

The compound $[Fe(\textit{meso}\text{-}Me_6[14]aneN_4)(CH_3CN)_2](CF_3SO_3)_2$ is only moderately sensitive to oxygen in the solid state but rapidly undergoes a complex sequence of reactions, involving isolatable iron(III) complexes, in the presence of oxygen in solution.[3,5] After a period of time iron(II) complexes containing dehydrogenated forms of the ligand can be isolated containing either three or four imine bonds in the five-membered chelate rings.[6] The complex should be stored in an inert atmosphere.[3] The infrared spectra of all complexes show a strong absorption near 3200 cm^{-1} due to the N—H function. The coordinated acetonitrile groups absorb in the 2230-2280 cm^{-1} region.

The acetonitrile complex is low-spin and six-coordinate. It gives only two bands in the visible region at 17.9 and 26.6 kK with intensities ($\epsilon < 50$) consistent with Laporte-forbidden, spin-allowed electronic transitions. Analysis of this spectrum after the method of Wentworth and Piper[7] leads to the results that $Dq(CH_3CN) \approx Dq(Me_6[14]aneN_4) \approx 2100\ cm^{-1}$ for low-spin iron(II).

References

1. W. L. Jolly, *Inorg. Synth.*, **11**, 116 (1968).

*See cautions concerning trifluoromethanesulfonic acid in Sections 1-A and 1-D.

†Because of the sensitivity of the solution toward oxygen, care should be taken if the reaction is performed on the bench top. An inert-atmosphere glove box would circumvent such difficulties.

2. J. J. Eisch and R. B. King (eds.), *Organometallic Chemistry*, Vol. 1, Academic Press, Inc., New York, 1965, p. 55.
3. J. C. Dabrowiak, P. H. Merrell, and D. H. Busch, *Inorg. Chem.*, **11**, 1979 (1972).
4. R. C. Dickinson and G. R. Long, *Chem. Eng. News*, **July 1970**, 6.
5. V. L. Goedken and D. H. Busch, *J. Am. Chem. Soc.*, **94**, 7355 (1972).
6. J. C. Dabrowiak, F. V. Lovecchio, V. L. Goedken, and D. H. Busch, *J. Am. Chem. Soc.*, **94**, 5502 (1972).
7. R. A. D. Wentworth and T. S. Piper, *Inorg. Chem.*, **4**, 709 (1965).

3. 2,12-DIMETHYL-3,7,11,17-TETRAAZABICYCLO[11.3.1]-HEPTADECA-1(17),2,11,13,15-PENTAENE (2,6-Me_2-2′,6′:3,5-Pyo[14]-1,3,6-triene-1,4,7,11-N_4) COMPLEXES

Submitted by A. MARTIN TAIT* and DARYLE H. BUSCH*
Checked by J. WILSHIRE† and A. B. P. LEVER†

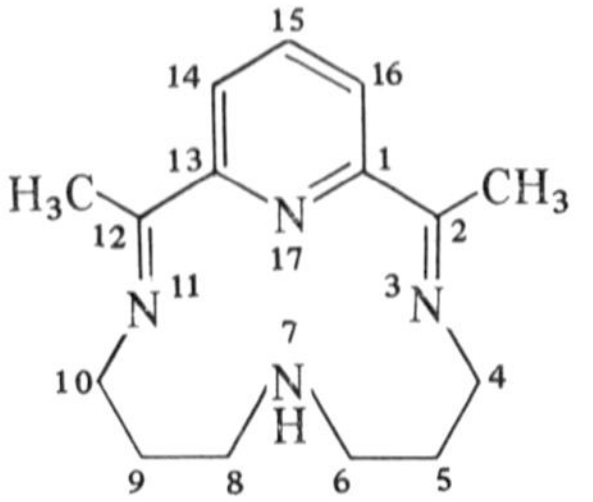

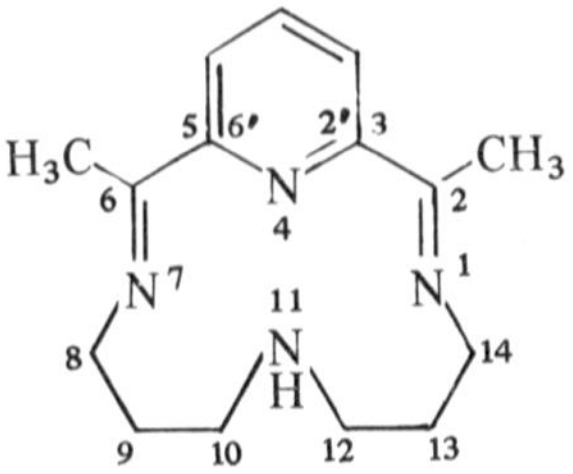

Numbering for abbreviation

In some of the earliest experiments involving *in situ* macrocyclic ligand synthesis, it was shown that the reaction of 2,6-diacetylpyridine with certain polyamines in the presence of metal ions leads to the preparation of new macrocyclic complexes.[1] As is often the case with Schiff-base condensations of the type, the addition of a small amount of acid catalyzes the reaction. Thus treatment of 2,6-diacetylpyridine with 3,3′-diaminodipropylamine [*N*-(3-aminopropyl)-1,3-propanediamine] in the presence of nickel(II) leads to the isolation of nickel complexes of the macrocyclic ligand (Me_2-Pyo[14]trieneN_4).[2-4]

*Department of Chemistry The Ohio State University, Columbus, OH 43210.
†Chemistry Department, York University, Downsview, Ontario M3J 1P3 Canada.

A. (2,12-DIMETHYL-3,7,11,17-TETRAAZABICYCLO[11.3.1]HEPTADECA-1(17),2,11,13,15-PENTAENE)NICKEL(II) PERCHLORATE

$$NiCl_2 \cdot 6H_2O + 2,6\text{-}(CH_3\underset{\underset{O}{\|}}{C}-)_2C_5H_3N + NH_2(CH_2)_3NH(CH_2)_3NH_2 \longrightarrow$$

$$8H_2O + [Ni(Me_2Pyo[14]trieneN_4)(H_2O)_2](ClO_4)_2]$$

$$[Ni(Me_2\text{-}Pyo[14]trieneN_4)(H_2O)_2](ClO_4)_2 \xrightarrow[\text{vacuum}]{P_4O_{10}}$$

$$[Ni(Me_2\text{-}Pyo[14]trieneN_4)](ClO_4)_2$$

Procedure

■ **Caution.** *See caution concerning perchlorates in Section 1-B.*

2,6-Diacetylpyridine (13 g, 0.08 mole) that has been recrystallized from petroleum ether (30-60° boiling range, 20 g/400 mL solvent) is dissolved in 160 mL of ethanol in a 1-L beaker. Nickel(II) chloride hexahydrate (19 g, 0.08 mole) dissolved in 240 mL of water is added and the solution is heated to 65° with stirring. 3,3′-Diaminodipropyl amine (10.5 g, 0.08 mole) is then added, followed by 5 mL of acetic acid to clarify the solution. The reaction mixture is heated for 6 hours at 65° and the solution volume is reduced to about 150 mL on a rotary evaporator. The solution is filtered, and concentrated aqueous sodium perchlorate (25 g, 0.2 mole, in 50 mL of water) is added to precipitate the crude product. The product is collected by filtration, washed with ethanol and dissolved in 300-400 mL of warm (65°) water. The solution is filtered hot, and 40 mL of 60% perchloric acid is added. The solution is cooled slowly and the grey-brown needles of the hydrated product $[Ni(Me_2\text{-}Pyo[14]trieneN_4)(H_2O)_2]$-$(ClO_4)_2$ are collected by filtration and washed with ethanol and diethyl ether. The crystals are dried *in vacuo* over P_4O_{10} to yield the brick-red anhydrous product.* The yield is 24 g (58%). *Anal.* Calcd. for $C_{15}H_{22}N_4Cl_2O_8Ni$: C, 34.91; H, 4.27; N, 10.86; Cl, 13.75. Found: C, 35.36; H, 4.41; N, 10.89; Cl, 14.06.

*Addition of lithium chloride or sodium thiocyanate to an acetone solution of the perchlorate salt causes an immediate precipitation of $[Ni(Me_2\text{-}Pyo[14]trieneN_4)Cl_2]$ or $[Ni(Me_2\text{-}Pyo[14]trieneN_4)(NCS)_2]$ which are olive-green and brown, respectively. Both of these products can be recrystallized from chloroform. The dichloro complex shows a tendency to hydrate on exposure to the atmosphere. Addition of a concentrated aqueous solution of sodium bromide to an aqueous solution of $[Ni(Me_2\text{-}Pyo[14]trieneN_4)](ClO_4)_2$ causes the precipitation of the complex $[Ni(Me_2\text{-}Pyo[14]trieneN_4)Br]ClO_4 \cdot H_2O$. Addition of sodium bromide to the dichloro complex in water produces the complex $[Ni(Me_2\text{-}Pyo[14]trieneN_4)Br]Br \cdot H_2O$. The bromo complexes are black.

Properties

The complex [Ni(Me_2-Pyo[14]trieneN$_4$)]$(ClO_4)_2$ is diamagnetic and square planar.[2] The infrared spectrum shows the N—H stretching mode at 3195 cm^{-1} and the C=N stretching frequency at 1578 cm^{-1}. The pyridine modes occur at 1570-1600 and at 820 cm^{-1}.

The NMR spectrum, in trifluoroacetic acid, shows the methyl singlet at 2.53 ppm. The visible spectrum shows bands at 17.9, 24.2 and 28.6 kK. The dichloro and dithiocyanato salts are high-spin ($\mu_{eff} \approx$ 3.15 BM) and six-coordinate.[2] Analysis[5] of the solid-state spectrum of the dithiocyanato complex, which shows *d-d* bands at 10.8, 18.9, and 25.4 kK, leads to a value for the ligand field strength (Dq^{xy}) of 1866 cm^{-1}, indicating that Me_2-Pyo[14]trieneN$_4$ is a very strong ligand toward nickel(II). The bromo complexes [Ni(Me_2-Pyo[14]-trieneN$_4$)Br]X·H_2O, where X = Br or ClO_4, are unusual in that they are diamagnetic and five-coordinate. An X-ray analysis[6] has shown that the macrocyclic molecules in Ni(Me_2-Pyo[14]trieneN$_4$)Br_2·H_2O are connected by N—H· · ·Br—Ni linkages so that the nickel atom has a distorted square-pyramidal stereochemistry with the bromide ligand in the axial position. The water molecule is strongly held between the coordinated and free bromide ions. The complex [Ni(Me_2-Pyo[14]trieneN$_4$)]$(ClO_4)_2$ has been hydrogenated with platinum oxide catalyst to yield two new isomeric complexes containing three saturated amine donor atoms and one pyridine nitrogen donor atom.[2,3,7]

B. BROMO(2,12-DIMETHYL-3,7,11,17-TETRAAZABICYCLO[11.3.1]-HEPTADECA-1(17),2,11,13,15-PENTAENE)COBALT(II) BROMIDE MONOHYDRATE

$$CoBr_2 + 2,6\text{-}(CH_3\underset{\underset{O}{\|}}{C})_2C_5H_3N + NH_2(CH_2)_3NH(CH_2)_3NH_2 \xrightarrow[N_2]{H^+}$$

$$[Co(Me_2\text{-Pyo}[14]\text{trieneN}_4)Br]Br\cdot H_2O + H_2O$$

Procedure

This procedure closely follows that for the preparation of [Ni(Me_2-Pyo[14]-trieneN$_4$)]$(ClO_4)_2$ [Sec. 3-A]. 2,6-Diacetylpyridine (16.3 g, 0.1 mole) that has been recrystallized from petroleum ether (30-60° boiling range, 20 g/400 mL solvent) is dissolved in 150 mL of hot degassed ethanol in a 500-mL flask equipped with a reflux condenser, dropping funnel, nitrogen inlet, and magnetic stirring bar. Water (100 mL) is then added, followed by 21.9 g (0.1 mole) of anhydrous cobalt(II) bromide (or the equivalent amount of the hexahydrate).* The solution is heated to 65-75° and 13.1 g (0.1 mole) of 3,3′-diaminodipropylamine is added from a dropping funnel with stirring, followed by 4 mL of glacial acetic acid. The dark-red solution is stirred at 65-75° for 4-6 hours under nitrogen. The solution volume is reduced to about 170 mL under vacuum with stirring and using mild heat. The solution is filtered under a blanket of nitrogen and 50 mL of a concentrated aqueous solution of lithium bromide (25 g/50 mL) is added, and the stoppered flask is set aside to cool to room temperature. The product is collected on a fritted-glass filter under a blanket of nitrogen and washed with a small amount of ethanol. This product is recrystallized from hot ethanol with the addition of lithium bromide solution (25 g in 50 mL water). The black needles of the monohydrate that form are filtered from the solution, washed with ethanol, and dried *in vacuo* over P_4O_{10}. The yield is 30 g (61%). *Anal.* Calcd. for $C_{15}H_{24}N_4Br_2OCo$: C, 36.38; H, 4.85; N, 11.32; Br, 32.30. Found: C, 36.60; H, 4.78; N, 11.36; Br, 32.50.

Properties

The dibromo and diisothiocyanato complexes are low-spin (μ_{eff} = 2.0-2.1 BM) and have been formulated as being five-coordinate with a trigonal bipyramidal geometry.[8] Both in the solid state and in solution, the electronic spectra show a weak band in the near-infrared region at about 9.0 kK, a series of bands near 14.0, 16.5, and 17.8 kK, and some more-intense bands in the ultraviolet region. The complexes behave as one-to-one electrolytes in methanol, nitromethane, and *N,N*-dimethylformamide. Their infrared spectra show N—H stretching frequencies in the 3145-3224 cm^{-1} region. A band near 1575 cm^{-1} has been assigned to the azomethine linkage, while bands near 1570, 1470, 1425, and 1205 cm^{-1} are assigned to the pyridine ring modes.

The cobalt(I) complexes of Me_2-Pyo[14]trieneN_4 react with alkyl halides[9-11] to produce cobalt(III) complexes of the type $[RCo(Me_2\text{-Pyo}[14]\text{triene}N_4)X]^+$ and $[R_2Co(Me_2\text{-Pyo}[14]\text{triene}N_4)]X$ containing cobalt-carbon bonds. The azomethine linkages of Me_2-Pyo[14]trieneN_4 are easily hydrogenated in methanol with platinum oxide catalyst,[8] yielding complexes in which the former azomethine linkages are now fully saturated.[12,13] Metathetical reactions on the complexes, $[Co(Me_2\text{-Pyo}[14]\text{triene}N_4)Br]Br{\cdot}H_2O$ and $[Co(Me_2\text{-Pyo}[14]\text{triene}N_4)(H_2O)](ClO_4)_2$, yield complexes of the type $[Co(Me_2\text{-Pyo}[14]\text{triene}N_4)X]X{\cdot}nH_2O$, where X = Cl^- or I^-, n = 0, 1, and $[Co(Me_2\text{-Pyo}[14]\text{triene}N_4)X](Y)_n$, where X = NH_3, Py, Br; Y = ClO_4^-, PF_6^-, $[B(C_6H_5)_4]^-$; n = 1, 2.

*Use of cobalt(II) thiocyanate or cobalt(II) nitrate hexahydrate in place of cobalt bromide leads to the isolation of $[Co(Me_2\text{-Pyo}[14]\text{triene}N_4)(NCS)]NCS$ or $[Co(Me_2\text{-Pyo}[14]\text{triene}N_4)NO_3]NO_3{\cdot}5H_2O$, respectively.

C. DIBROMO[2,12-DIMETHYL-3,7,11,17-TETRAAZABICYCLO[11.3.1]-HEPTADECA-1(17),2,11,13,15-PENTAENE)COBALT(III) BROMIDE MONOHYDRATE

$$CoBr_2 + 2,6\text{-}(CH_3\underset{\underset{O}{\|}}{C})_2C_5H_3N + NH_2(CH_2)_3NH(CH_2)_3NH_2 \xrightarrow[H^+]{N_2}$$

$$\xrightarrow[HBr]{O_2} [Co(Me_2\text{-}Pyo[14]trieneN_4)Br_2]Br\cdot H_2O + H_2O$$

*Procedure**

The procedure is identical to that used to prepare the analogous cobalt(II) complex (Sec. 3-B) except that after heating the reaction mixture for 6 hours at 65-75° under nitrogen, the solution is cooled and 16.9 g (0.1 mole) of 47-49% hydrobromic acid is added. The solution is aerated for 6 hours until precipitation of the green product is complete. The product is collected by filtration, washed with ethanol, and recrystallized from several hundred milliliters of hot 5% aqueous hydrobromic acid. The lime-green crystals are washed with ethanol and dried *in vacuo* over P_4O_{10}. The yield is 40 g (70%). *Anal.* Calcd. for $C_{15}H_{24}N_4Br_3OCo$: C, 31.32; H, 4.18; N, 9.74; Br, 41.72. Found: C, 31.50; H, 4.14; N, 9.81; Br, 41.56.

Properties

The cobalt(III) complexes are low-spin and six-coordinate, with the macrocyclic ligand arranged in a planar fashion.[13,14] They are univalent electrolytes in methanol and in nitromethane. Their infrared spectra are very similar to the analogous cobalt(II) complexes. Visible spectra show transitions at 20.6 kK and in the 14.7-16.6 and 25.3-28.5 kK regions. Analysis[13-15] of the spectra leads to a value for Dq^{xy} of 2820 cm^{-1} for the bromo complex, indicating that the ligand Me_2-Pyo[14]trieneN$_4$ binds very strongly to cobalt(III). The NMR spectra, recorded in acidic D_2O show a sharp methyl singlet near 3.50 ppm. The pyridine protons occur as a typical AB_2 pattern near 9.00 ppm. The complexes are sensitive to base hydrolysis and are inert towards hydrogenation of the ligand with

*The analogous complex [Co(Me_2-Pyo[14]trieneN$_4$)Cl_2]Cl·H_2O is prepared in an identical manner using cobalt(II) chloride hexahydrate and hydrochloric acid instead of cobalt(II) bromide and hydrobromic acid. Use of cobalt(II) nitrate and nitric acid yields, after the concentrated reaction mixture is allowed to stand for several days, brown crystalline [Co(Me_2-Pyo[14]trieneN$_4$)$(NO_3)_2$]$NO_3\cdot 2.5H_2O$.

platinum oxide catalyst.[13,14] Other complexes containing monodentate axial ligands (N_3^-, NCS^-, NO_2^-, CN^-, I^-, ClO_4^-) or bidentate ligands [oxalate, acetylacetone (2,4-pentanedione), which produce unusual octahedral *cis*-complexes in which the macrocyclic ligand is folded] have been prepared and characterized.[14]

References

1. J. D. Curry and D. H. Busch, *J. Am. Chem. Soc.*, **86**, 592 (1964).
2. J. L. Karn and D. H. Busch, *Nature*, **211**, 160 (1966).
3. J. L. Karn and D. H. Busch, *Inorg. Chem.*, 8, 1149 (1969).
4. R. L. Rich and G. L. Stucky, *Inorg. Nucl. Chem. Lett.*, **1**, 85 (1965).
5. C. R. Sperati, Ph.D. Thesis, The Ohio State University, 1971.
6. E. B. Fleischer and S. W. Hawkinson, *Inorg. Chem.*, 7, 2312 (1968).
7. E. Ochiai and D. H. Busch, *Inorg. Chem.*, 8, 1798 (1969).
8. K. L. Long and D. H. Busch, *Inorg. Chem.*, **9**, 505 (1970).
9. E. Ochiai and D. H. Busch, *Chem. Commun.*, **1968**, 905.
10. E. Ochiai, K. M. Long, C. R. Sperati, and D. H. Busch, *J. Am. Chem. Soc.*, **91**, 3201 (1969).
11. K. Farmery and D. H. Busch, *Chem. Commun.*, **1970**, 1091; *Inorg. Chem.*, **11**, 2901 (1972).
12. E. Ochiai and D. H. Busch, *Inorg. Chem.*, 8, 1474 (1969).
13. K. M. Long, Ph.D. Thesis, The Ohio State University, 1967.
14. K. M. Long and D. H. Busch, *J. Coord. Chem.*, **4**, 113 (1974).
15. S. C. Jackels, K. Farmery, E. K. Barefield, N. J. Rose, and D. H. Busch, *Inorg. Chem.*, **11**, 2893 (1972).

4. 2,3,9,10-TETRAMETHYL-1,4,8,11-TETRAAZACYCLOTETRADECA-1,3,8,10-TETRAENE (2,3,9,10-Me_4[14]-1,3,8,10-tetraene-1,4,8,11-N_4) COMPLEXES

Submitted by A. MARTIN TAIT* and D. H. BUSCH*
Checked by M. A. KHALIFA† and J. CRAGEL, JR.†

*Department of Chemistry, The Ohio State University, Columbus, OH 43210.
†Department of Chemistry, University of Pittsburgh, Pittsburgh, PA 15260.

The most obvious and most often attempted macrocyclization reaction, by way of the Schiff-base, involves the condensation of α-diketones with 1,2- or 1,3-diamines. The resulting macrocyclic ligands, containing the *in situ* generated α-diimine group, are of interest because of the special properties often associated with the α-diimine linkage.[1] The first macrocyclic ligand to be prepared with such a grouping was 2,3,9,10-$Me_4[14]$-1,3,8,10-tetraene-1,4,8,11-N_4.[2] This ligand, which contains two α-diimine linkages, is prepared in the presence of nickel(II), cobalt(II), or iron(II) by the condensation of 2,3-butanedione with 1,3-propanediamine in the presence of a suitable acid.[2-5]

A. (2,3,9,10-TETRAMETHYL-1,4,8,11-TETRAAZACYCLOTETRADECA-1,3,8,10-TETRAENE)NICKEL(II) PERCHLORATE

$$Ni(OOCCH_3)_2 \cdot 4H_2O + 2NH_2(CH_2)_3NH_2 \cdot HCl + 2CH_3\underset{\underset{O}{\|}}{C}-\underset{\underset{O}{\|}}{C}CH_3 \xrightarrow[HCl]{ZnCl_2}$$

$$[Ni(Me_4[14]\text{-}1,3,8,10\text{-tetraene}N_4)][ZnCl_4] + 8H_2O + 2CH_3COOH$$

$$(Ni(Me_4[14]\text{-}1,3,8,10\text{-tetraene}N_4)][ZnCl_4] + 4AgNO_3 \longrightarrow$$

$$[Ni(Me_4[14]\text{-}1,3,8,10\text{-tetraene}N_4](NO_3)_2 + 4AgCl + Zn(NO_3)_2$$

$$[Ni(Me_4[14]\text{-}1,3,8,10\text{-tetraene}N_4)](NO_3)_2 + 2NaClO_4 \longrightarrow$$

$$[Ni(Me_4[14]\text{-}1,3,8,10\text{-tetraene}N_4)](ClO_4)_2 + 2NaNO_3$$

Procedure

■ **Caution.** *See caution concerning perchlorates in Section 1-B.*

To 1 L of methanol in a 2-L beaker containing 15 g (0.2 mole) of 1,3-propanediamine is added 19.75 g (0.2 mole) of concentrated (37%) hydrochloric acid from a dropping funnel. The solution is cooled to 5° and 17.2 g (0.2 mole) of 2,3-butanedione (biacetyl) is added to the mixture in an efficient fume hood. The solution is stirred for 30 minutes and then allowed to stand at room temperature. After about 20 minutes, 24.9 g (0.1 mole) of nickel(II) acetate tetrahydrate is added to the orange solution. The solution darkens to a red-brown and after 4 hours of stirring 19.7 g (0.2 mole) of concentrated hydrochloric acid is added, followed by 13.6 g (0.1 mole) of zinc chloride. The dark brown-red solid form of $[Ni(Me_4[14]$-1,3,8,10-tetraene$N_4)][ZnCl_4]$ precipitates immediately* and is removed by filtration, washed with cold ethanol and diethyl ether,

*If the tetrachlorozincate salt does not precipitate immediately upon addition of zinc chloride, the reaction mixture should be stirred overnight, followed, if necessary, by the addition of small quantities of tetrahydrofuran to precipitate the desired product.

and dried *in vacuo* over P_4O_{10}. The yield is 18 g (35%). *Anal.* Calcd. for $C_{14}H_{24}N_4Cl_4ZnNi$: C, 32.69; H, 4.67; N, 10.90; Cl, 27.60. Found: C, 32.72, H, 4.59; N, 10.99; Cl, 27.41.

This tetrachlorozinate salt (5.1 g, 0.01 mole) is dissolved in 60 mL of water, and 6.8 g (0.04 mole) of silver nitrate is added. The mixture is stirred for 30 minutes, after which it is filtered to remove the precipitated silver chloride. Sodium perchlorate (2.45 g, 0.02 mole) is added to the red-orange filtrate and the solution is taken to dryness on a rotary evaporator*. The solids are washed into a filter funnel with small amounts of acetone and water. The yellow $[Ni(Me_4[14]\text{-}1,3,8,10\text{-tetraene}N_4)](ClO_4)_2$ complex† is dried *in vacuo* over P_4O_{10}. The yield is 3 g (59%). *Anal.* Calcd. for $C_{14}H_{24}N_4Cl_2O_8Ni$: C, 33.23; H, 4.75; N, 11.07. Found; C, 33.15; H, 5.02; N, 10.96.

B. BIS(ISOTHIOCYANATO)(2,3,9,10-TETRAMETHYL-1,4,8,11-TETRA-AZACYCLOTETRADECA-1,3,8,10-TETRAENE)NICKEL(II)

$$[Ni(Me_4[14]\text{-}1,3,8,10\text{-tetraene}N_4)](ClO_4)_2 + 2KNCS \longrightarrow [Ni(Me_4[14]\text{-}1,3,8,10\text{-tetraene}N_4)(NCS)_2] + 2KClO_4$$

Procedure‡

■ **Caution.** *See caution concerning perchlorates in Section 1-B.*

One hundredth mole (5.1 g) of $[Ni(Me_4[14]\text{-}1,3,8,10\text{-tetraene}N_4)](ClO_4)_2$ is slurried in 100 mL of absolute ethanol, and 2 mL of a saturated aqueous solution of potassium thiocyanate is added. The solution is refluxed for 30 minutes, and the precipitated potassium perchlorate is removed from the cold solution by filtration. The red-purple filtrate is evaporated to dryness on a rotary evaporator, and 100 mL of chloroform is added. The solution is filtered and diethyl ether is added to precipitate the complex. The red-brown solid is washed with diethyl ether and dried *in vacuo* over P_4O_{10}. The sample is slightly moisture sensitive. The yield is 3.5 g (83%). *Anal.* Calcd. for $C_{16}H_{24}N_6S_2Ni$: C, 45.41; H, 5.67; N, 19.87; S, 15.16. Found: C, 45.08; H, 5.59; N, 19.72; S, 14.71.

*The addition of sodium hexafluorophosphate to the aqueous solution of $[Ni(Me_4[14]\text{-}1.3.8.10\text{-tetraene}N_4)](NO_3)_2$ in the above procedure leads to the isolation of the hexafluorophosphate salt of the complex.

†The perchlorate complex can also be obtained by addition of excess sodium perchlorate to a hot saturated aqueous solution of the tetrachlorozincate salt. The solution is refluxed for 30 minutes and then cooled to room temperature. The material that precipitates is recrystallized several times from water containing sodium perchlorate.

‡The complex can also be prepared by refluxing an aqueous solution of excess potassium thiocyanate and $[Ni(Me_4[14]\text{-}1,3,8,10\text{-tetraene}N_4)][ZnCl_4]$ for 1 hour and extracting the dry solids into chloroform.

Properties

The complex [Ni(Me_4[14]-1,3,8,10-tetraeneN_4)$(NCS)_2$] is paramagnetic (μ_{eff} = 3.08 BM) and six-coordinate[2,5] and behaves as a nonelectrolyte in nitromethane and acetonitrile solutions. The visible spectrum of a solid sample at liquid-nitrogen temperature shows bands at 11.9, 17.7, 19.0, and 23.0 kK leading to a value of Dq^{xy} = 1767 cm^{-1}, making the macrocycle one of the strongest cyclic ligands studied for nickel(II).[5] All the other salts (ClO_4^-, PF_6^-, $ZnCl_4^{2-}$) are diamagnetic and square planar, with intense bands in their visible spectra occurring near 25.0 kK. The NMR spectrum of the perchlorate salt in nitromethane shows a single methyl resonance at 2.40 ppm.[3] The infrared spectra of all these nickel complexes show weak imine C=N stretches in the 1600-1550 cm^{-1} region, with a strong sharp band near 1210 cm^{-1} being assigned to the N=C−C=N function. This band disappears on complete hydrogenation of the ligand on nickel[3] (with Raney nickel and H_2) to produce the fully saturated complex [Ni(2,3,9,10-Me_4-[14]ane-1,4,8,11-N_4)]$^{2+}$.

C. DIBROMO(2,3,9,10-TETRAMETHYL-1,4,8,11-TETRAAZACYCLOTETRADECA-1,3,8,10-TETRAENE)COBALT(III) BROMIDE

$$Co(OOCCH_3)_2 \cdot 4H_2O + 2NH_2(CH_2)_3NH_2 \cdot HBr + 2CH_3\underset{\displaystyle O}{\underset{\|}{C}}-\underset{\displaystyle O}{\underset{\|}{C}}-CH_3 \xrightarrow[3\ hr]{N_2}$$

$$\xrightarrow[HBr]{O_2} [Co(Me_4[14]\text{-}1,3,8,10\text{-tetraene}N_4)Br_2]Br + 2CH_3COOH + 8H_2O + H^+$$

*Procedure**

Aqueous hydrobromic acid (32.6 g, 0.2 mole of 48% HBr) is added slowly from a dropping funnel to a solution of 1,3-propanediamine (15 g, 0.2 mole) in 600 mL of degassed methanol.[4] 2,3-Butanedione (biacetyl, 17.2 g, 0.2 mole) is added under nitrogen from a dropping funnel over a 30-minute period† in a well-ventilated hood and the mixture is stirred. The solution slowly changes color from yellow to orange-red. After addition of the 2,3-butanedione is completed, hydrated cobalt(II) acetate (24.9 g, 0.1 mole) is added, and the solution is stirred under nitrogen for 3 hours at room temperature. To the resultant deep-purple solution is added about 70 mL of 48% aqueous hydrobromic acid, and air

*Use of hydrochloric or hydroiodic acids in place of hydrobromic acid leads to the analogous dichloro and diiodo complexes.

†Yields decrease significantly if the addition time is greater than 30 minutes. The 2,3-butanedione must be added dropwise, and the solution color at the end of the addition should be orange-red as opposed to red.

is bubbled through the solution overnight. The resulting lime-green precipitate is washed with 5% acidic (HBr) methanol and recrystallized from hot 5% aqueous hydrobromic acid to yield dark-green crystals which are dried *in vacuo* over P_4O_{10}. The yield is 14 g (26%). *Anal.* Calcd. for $C_{14}H_{24}N_4Br_3Co$: C, 30.73; H, 4.39; N, 10.24; Br, 43.83. Found: C, 30.72; H, 4.46; N, 10.40; Br, 43.56.

Properties

The complex $[Co(Me_4[14]\text{-}1,3,8,10\text{-tetraene}N_4)Br_2]Br$ is diamagnetic and six-coordinate.[4] The visible spectrum shows bands near 16.9, 25.5, and 32.7 kK leading to a value of Dq^{xy} = 2860 cm^{-1}. The NMR of the dibromo complex shows a methyl resonance at 3.35 ppm in dimethyl sulfoxide. Metathetical reactions performed in methanol or water or their mixture using sodium or potassium salts produce a great variety of tetragonally distorted complexes of the type $[Co(Me_4[14]\text{-}1,3,8,10\text{-tetraene}N_4)X_2]Y$ where X = N_3^-, NCS^-, CH_3COO^-, $HCOO^-$, NO_2^-, CN^-, and Y = ClO_4^-, PF_6^-, $[B(C_6H_5)_4]^-$.

The infrared spectra of the cobalt complexes are similar to those observed for the nickel(II) complexes. The cobalt(III) complexes of $2,3,9,10\text{-}Me_4[14]\text{-}1,3,8,10\text{-tetraene-}1,4,8,11\text{-}N_4$ can be converted to air-stable, but light-sensitive, complexes containing one or two alkyl groups bonded to the cobalt atom.[6,7] The imine functions of $2,3,9,10\text{-}Me_4[14]\text{-}1,3,8,10\text{-tetraene-}1,4,8,11\text{-}N_4$ can be hydrogenated on cobalt with sodium tetrahydroborate(1−)[7] or with hypophosphorous acid[8] to produce cobalt(III) complexes of $2,3,9,10\text{-}Me_4[14]\text{ane-}1,4,8,11\text{-}N_4$ and $2,3,9,10\text{-}Me_4[14]\text{-}1,8\text{-diene-}1,4,8,11\text{-}N_4$, respectively.

References

1. L. F. Lindoy and S. E. Livingstone, *Coord. Chem. Rev.*, **2**, 173 (1967).
2. D. A. Baldwin and N. J. Rose, Abstracts of the 157th Meeting of the American Chemical Society, Paper No. 20, Minneapolis, Minn., 1969: D. A. Baldwin, R. M. Pfeiffer, D. W. Reichgott, and N. J. Rose, *J. Am. Chem. Soc.*, **95**, 5152 (1973).
3. E. K. Barefield, Ph.D. Thesis, The Ohio State University, 1969.
4. S. C. Jackels, K. Farmery, E. K. Barefield, N. J. Rose, and D. H. Busch, *Inorg. Chem.*, **11**, 2893 (1972).
5. C. R. Sperati, Ph.D. Thesis, The Ohio State University, 1971.
6. K. Farmery and D. H. Busch, *Chem. Commun.*, **1970**, 1041.
7. K. Farmery and D. H. Busch, *Inorg. Chem.*, **11**, 2901 (1972).
8. A. M. Tait and D. H. Busch, *Inorg. Chem.*, **16**, 966 (1977).

5. 2,3-DIMETHYL-1,4,8,11-TETRAAZACYCLOTETRADECA-1,3-DIENE (2,3-Me$_2$[14]-1,3-diene-1,4,8,11-N$_4$) COMPLEXES

Submitted by A. MARTIN TAIT* and DARYLE H. BUSCH*
Checked by J. CRAGEL, JR.†

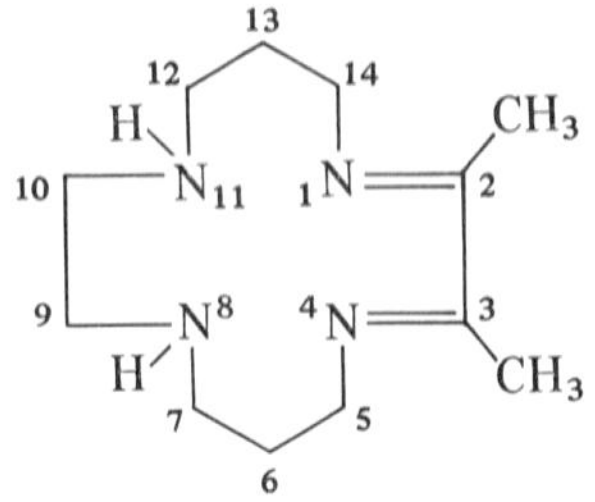

The condensation of *N,N'*-bis(3-aminopropyl)ethylenediamine (*N,N''*-ethylenebis[1,3-propanediamine]) as its acid salt with 2,3-butanedione (biacetyl) in the presence of cobalt(II) or nickel(II) acetate gives complexes of 2,3-Me$_2$[14]-1,3-diene-1,4,8,11-N$_4$ containing one α-diimine linkage.[1,2] Experiments have shown that the presence of H^+ ion determines whether or not a macrocyclic complex forms and that, in the presence of H^+, the time at which the metal acetate is added to the reaction mixture influences the yield of the complex.[2] Unlike the reaction between biacetyl and 1,3-propanediamine to form 2,3,9,10-Me$_4$[14]-1,3,8,10-tetraene-1,4,8,11-N$_4$ (Sec. 4), the condensation of biacetyl with *N,N'*-bis(3-aminopropyl)ethylenediamine is particularly sensitive to excess acetate so that the procedures given use the optimized conditions.

A. (2,3-DIMETHYL-1,4,8,11-TETRAAZACYCLOTETRADECA-1,3-DIENE)-NICKEL(II) TETRACHLOROZINCATE(2−)

$$Ni(OOCCH_3)_2 \cdot 4H_2O + NH_2(CH_2)_3NH(CH_2)_2NH(CH_2)_3NH_2 \cdot HCl +$$

$$CH_3-\underset{\underset{O}{\|}}{C}-\underset{\underset{O}{\|}}{C}-CH_3 \xrightarrow[HCl]{ZnCl_2}$$

$$[Ni(Me_2[14]\text{-}1,3\text{-}dieneN_4)][ZnCl_4] + 6H_2O + 2CH_3COOH$$

*Department of Chemistry, The Ohio State University, Columbus, OH 43210.
†Department of Chemistry, University of Pittsburgh, Pittsburgh, PA 15260

Preparation

To 1600 mL of methanol in a 3-L flask is added 58 g (0.33 mole) of *N,N'*-bis-(3-aminopropyl)ethylenediamine* and, dropwise, 32.8 g (0.33 mole) of concentrated hydrochloric acid. The solution is cooled to 5° and stirred while 28.7 g (0.33 mole) of 2,3-butanedione is added. After 30 minutes the flask is removed from the ice bath and allowed to stand at room temperature. After a further 20 minutes the solution is pale orange and 82.6 g (0.33 mole) of nickel(II) acetate tetrahydrate is added. The mixture is stirred for 4 hours and 35 mL of concentrated hydrochloric acid is added to the deep red-brown solution. Addition of 45.4 g (0.33 mole) of zinc chloride produces an immediate red-brown precipitate, which is collected by filtration and washed thoroughly wtih diethyl ether. The yield is 110-120 g (68-74%). *Anal.* Calcd. for $C_{12}H_{24}N_4Cl_4ZnNi$: C, 29.39; H, 4.90; N, 11.43; Cl, 28.95. Found: C, 29.25; H, 5.10; N, 11.30; Cl, 28.41.

Properties

The complex [Ni(2,3-Me_2[14]-1,3-diene-1,4,8,11-N_4)] [$ZnCl_4$] is square planar and low-spin.[1,3] The visible spectra show bands near 21.3 kK (characteristic of square planar nickel(II)), near 26.1 kK (due to the imine functions), and near 35.1 kK. The infrared spectra of all of the nickel complexes prepared show absorptions near 3195 and 1595 cm^{-1} assignable to the N—H stretching vibration and to the symmetric imine vibration, respectively.[1] A strong sharp band also occurs near 1210 cm^{-1} and is characteristic of the α-diimine function. The NMR spectrum of the perchlorate complex in nitromethane shows a methyl singlet at 2.33 ppm. The ligand can be hydrogenated on nickel(II) with Raney nickel and hydrogen to produce the fully saturated macrocyclic complex [Ni(2,3-Me_2[14]-ane-1,4,8,11-N_4]$^{2+}$.[1,3]

B. DIBROMO(2,3-DIMETHYL-1,4,8,11-TETRAAZACYCLOTETRADECA-1,3-DIENE)COBALT(III) PERCHLORATE

$$Co(OOCCH_3)_2 \cdot 4H_2O + NH_2(CH_2)_3NH(CH_2)_2NH(CH_2)_3NH_2 \cdot HBr +$$

**N,N'*-Bis(3-aminopropyl)ethylenediamine is prepared by the method of E. K. Barefield.[1] Six moles of 1,3-propanediamine is dissolved in 1.4 L of absolute ethanol and the solution is cooled to 5° in an ice bath. To this solution is added 0.75 mole of 1,2-dibromoethane from a dropping funnel with vigorous stirring. After the addition is complete, the reaction is heated to reflux temperature for 1½ hours. Potassium hydroxide, 150 g, is added and the mixture is refluxed for a further 1 hour. The reaction mixture is cooled to room temperature and filtered to remove the solids. The filtrate is evaporated to a sludge on a rotary evaporator and the semisolid is extracted several times with diethyl ether. The ether solution is evaporated until a viscous liquid remains. The liquid is distilled *in vacuo* (b.p. 138-148°/2 torr) and stored over potassium hydroxide pellets in a bottle protected from light. The yield is 64 g.

$$CH_3-\underset{\underset{O}{\|}}{C}-\underset{\underset{O}{\|}}{C}-CH_3 \xrightarrow[4\ hr]{N_2} \xrightarrow[NaClO_4]{O_2/HBr} 6H_2O + 2CH_3COOH + Na^+$$

$$+ [Co(Me_2[14]\text{-}1,3\text{-diene}N_4)Br_2]ClO_4$$

*Procedure**

■ **Caution.** *See caution concerning perchlorates in Section 1-B.*

Aqueous hydrobromic acid (3.3 g, 0.02 mole of 48% HBr) is added slowly from a dropping funnel to a solution of *N,N'*-bis(3-aminopropyl)ethylenediamine (3.5 g, 0.02 mole) in 60 mL of methanol under nitrogen. 2,3-Butanedione (1.7 g, 0.02 mole) is added dropwise, and the mixture is stirred for 30 min. Hydrated cobalt(II) acetate (2.5 g, 0.01 mole) is added to the yellow-brown solution and the mixture is stirred under nitrogen for 4 hr. Excess hydrobromic acid (5 mL) is added to the purple solution, which is aerated until the solution turns green. Excess sodium perchlorate (3 g) is added with stirring and the resultant bright-green solid is removed by filtration, washed with acidic methanol (5% HBr), and dried *in vacuo* over P_4O_{10}. The yield is 2 g (37% based on cobalt(II) acetate). *Anal.* Calcd. for $C_{12}H_{24}N_4Br_2ClO_4Co$: C, 26.56; H, 4.43; N, 10.33; Br, 29.47. Found: C, 26.50; H, 4.48; N, 10.15; Br, 29.28.

Properties

The cobalt(III) complex [Co(2,3-Me_2[14]-1,3-diene-1,4,8,11-N_4)Br_2]Br is six-coordinate and diamagnetic. Analysis[2] of the visible spectrum (absorptions occur near 15.8 and 26.2 kK) leads to a value for Dq^{xy} of 2630 cm^{-1}. The NMR spectrum in dimethyl sulfoxide shows a methyl singlet at 3.32 ppm. The infrared spectrum is very similar to that given for the nickel(II) complexes. A wide variety of cobalt(III) complexes has been prepared by metathetical reactions on the dibromo and dichloro complexes. The imine functions can be hydrogenated, producing cobalt(III) complexes of 2,3-Me_2[14]ane-1,4,8,11-N_4[2] or 2,3-Me_2-[14]-l-ene-1,4,8,11-N_4.[4]

References

1. E. K. Barefield, Ph.D. Thesis, The Ohio State University, 1969.
2. S. C. Jackels, K. Farmery, E. K. Barefield, N. J. Rose, and D. H. Busch, *Inorg. Chem.*, **11**, 2893 (1972).
3. C. R. Sperati, Ph.D. Thesis, The Ohio State University, 1971.
4. A. M. Tait and D. H. Busch, *Inorg. Chem.*, **16**, 966 (1977).

*Use of hydrochloric acid instead of hydrobromic acid in the procedure described above leads to the corresponding *trans*-dichloro complex.

6. TETRABENZO[*b,f,j,n*] [1,5,9,13] TETRAAZACYCLOHEXADECINE (2,3;6,7;10,11;14,15-Bzo_4[16]octaene-1,5,9,13-N_4 or Bzo_4[16]octaeneN_4) COMPLEXES

Submitted by A. MARTIN TAIT* and DARYLE H. BUSCH*
Checked by J. WILSHIRE† and A. B. P. LEVER†

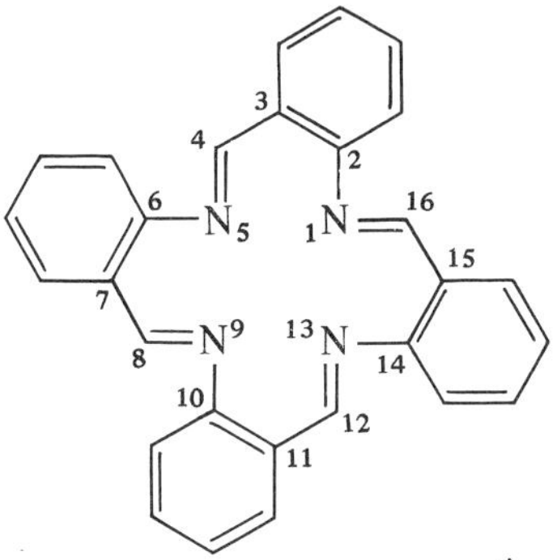

Numbering for abbreviation‡

It is known that *o*-aminobenzaldehyde condenses with itself to form a number of products, including a tricyclic trimer and a tricyclic tetramer.[1] In the presence of metal ions, however, the self-condensation of *o*-aminobenzaldehyde proceeds in a different manner and the metal complexes with macrocyclic Schiff-base ligands are formed.[2-11] Complexes containing either a closed 12-membered tridentate macrocyclic ligand, tribenzo[*b,f,j,*] [1,5,9]triazacyclododecine (2,3;6,7;10,11-Bzo_3[12]hexaene-1,5,9-N_3), containing 3 moles of *o*-aminobenzaldehyde, or a cyclic quadridentate ligand, tetrabenzo[*b,f,j,n*]-[1,5,9,13]tetraazacyclohexadecine [Bzo_4[16]octaeneN_4], have been isolated with nickel(II)[2-8] and cobalt(III)[9-10] ions. In the presence of iron(II),[11] copper(II)[4,6], and zinc(II)[12] ions, the complexes contain exclusively the quadridentate ligand Bzo_4[16]octaeneN_4. The results of an extensive study on these reactions have shown that the self-condensation of *o*-aminobenzaldehyde is governed by coordination template effects, since the products in the presence of metal ions are variable and different from those formed in their absence. All the Bzo_4[16]octaeneN_4 complexes show remarkable stability to concentrated acids but readily react with nucleophiles (e.g., ethoxide ion), which add to the coordinated imine groups, to form new macrocyclic complexes.[13,14]

*Department of Chemistry, The Ohio State University, Columbus, OH 43210.
†Chemistry Department York University, Downsview, Ontario, M3J 1P3 Canada.
‡Since the benzene rings are located by letters in the name, only the locant numbers of the N (in square brackets since they do not refer to the total ring) are needed for the name.

A. (TETRABENZO[*b,f,j,n*][1,5,9,13]TETRAAZACYCLOHEXADECINE)-NICKEL(II) PERCHLORATE

$$Ni(NO_3)_2 \cdot 6H_2O + 4(NH_2C_6H_4CHO) + 2NaClO_4 \longrightarrow$$
$$[Ni(Bzo_4[16]octaeneN_4](ClO_4)_2 + [Ni(Bzo_3[12]hexaeneN_3)H_2O(ClO_4)_2]$$

Procedure

■ **Caution.** *See caution concerning perchlorates in Section 1-B.*

A solution of 2.9 g (0.024 mole) of freshly prepared *o*-aminobenzaldehyde* in 40 mL of absolute ethanol is heated to reflux with stirring. An ethanolic solution (30 mL) of 1.74 g (0.006 mole) of hydrated nickel(II) nitrate is added. The solution immediately turns from pale yellow to dark brown and after about 30 minutes an orange product appears. The solution is stirred and refluxed for a further 7 hours, cooled, and filtered. The isolated orange product is washed with ethanol and diethyl ether and is dried *in vacuo* over P_4O_{10}. The yield is 2.65 g. This orange material is dissolved in 250 mL of water at room temperature and a concentrated aqueous solution of sodium perchlorate (2.0 g, 0.016 mole in 10 mL of water) is added. A bright-red precipitate is produced immediately on stirring. The solution is filtered† and the product is washed with water and dried *in vacuo* over P_4O_{10}. The yield is 1.3 g (33%). *Anal.* Calcd. for $C_{28}H_{20}N_4Cl_2O_8Ni$: C, 50.18; H, 2.99; N, 8.36; Cl, 10.59. Found: C, 49.97; H, 3.29; N, 8.59; Cl, 10.57.

B. BIS(ISOTHIOCYANATO)(TETRABENZO[*b,f,j,n*][1,5,9,13]TETRA-AZACYCLOHEXADECINE)NICKEL(II)

$$[Ni(Bzo_4[16]octaeneN_4)](ClO_4)_2 + 2NaNCS \longrightarrow$$
$$[Ni(Bzo_4[16]octaeneN_4)(NCS)_2] + 2NaClO_4$$

Procedure

One gram (0.0015 mole) of $[Ni(Bzo_4[16]octaeneN_4)](ClO_4)_2$ is dissolved in 500 mL of water, and 10 mL of sodium thiocyanate (0.5 g, in 10 mL of water) is

*o-Aminobenzaldehyde is prepared from *o*-nitrobenzaldehyde (Aldrich Chemical Co.) by the method of Smith and Opie.[15] The white crystalline material should be used immediately but can be stored at 0° for short periods of time without decomposition.

†The filtrate remaining after the isolation of the red crystalline $[Ni(Bzo_4[16]octaeneN_4)]$-$(ClO_4)_2$ complex yields, on concentration of the solution, a small amount of a yellow complex of the tridentate ligand $Bzo_3[12]hexeneN_3$.[5]

added.* The complex precipitates immediately on stirring. The complex is isolated by filtration, washed with water, and dried *in vcauo* over P_4O_{10}. The yield is nearly quantitative. *Anal.* Calcd. for $C_{30}H_{20}N_6S_2Ni$: C, 61.35; H, 3.41; N, 41.31; S, 10.93. Found: C, 61.03; H, 3.51; N, 14.21; S, 10.80.

Properties

The red perchlorate complex is diamagnetic and the solid reflectance spectrum shows a band at 18.7 kK typical of square planar nickel(II).[3,4] The isothiocyanato complex is six-coordinate and high-spin (μ_{eff} = 3.21 BM)[3,4] and exhibits visible spectral bands at 10.5, 11.9, and 18.2 kK. The infrared spectra of both complexes exhibit bands near 1610, 1591, 1492, and 1448 cm^{-1} due to the ortho-disubstituted benzene moiety and a strong sharp band near 1570 cm^{-1} assignable to the imine functions. The square planar and octahedral structures have been verified by X-ray crystal structure analyses of the analogous $[Ni(Bzo_4[16]octaeneN_4)](BF_4)_2$ and $[Ni(Bzo_4[16]octaeneN_4)(H_2O)I]I$ complexes that show the $Bzo_4[16]octaeneN_4$ ligand is not planar but assumes a decided "saddle" shape,[16] with the four nitrogen donors essentially coplanar.

The complexes are extremely stable to boiling concentrated mineral acids[4] but are sensitive to base (alkoxides or amines) producing, by way of nucleophilic attack on two of the azomethine linkages, neutral four-coordinate[13] or five-coordinate complexes.[13,14]

Hydrogenation of the imine functions of the ligand has been achieved either with platinum oxide catalyst[17] or by electrochemical means[18,19] to produce a new series of fully saturated (excluding the benzene rings) nickel(II) complexes as well as species that appear to contain aromatic dianionic ligands.

C. (TETRABENZO[*b,f,j,n*][1,5,9,13]TETRAAZACYCLOHEXADECINE)-COPPER(II) NITRATE

$$Cu(NO_3)_2 \cdot 3H_2O + 4(NH_2C_6H_4CHO) \longrightarrow [Cu(Bzo_4[16]octaeneN_4)](NO_3)_2 + 7H_2O$$

*Addition of the appropriate sodium salt and a few milliliters of the corresponding acid, if available, to an aqueous solution of $[Ni(Bzo_4[16]octaeneN_4)](ClO_4)_2$ leads to the formation of complexes formulated as square planar; $[Ni(Bzo_4[16]octaeneN_4)]X_2$ where X = BF_4^-, and $[B(C_6H_5)_4]^-$ or octahedral $[Ni(Bzo_4[16]octaeneN_4)X_2]\cdot nH_2O$ where X = Cl^-, Br^-, I^-, NO_3^- and n = 0, 1. All complexes precipitate immediately from the reaction mixture, with the exception of the iodo complex, which crystallizes on standing for 2 days, and the nitrato complex, which crystallizes on reducing the solution volume to a few milliliters.

Procedure

A solution of 2.9 g (0.024 mole) of freshly prepared *o*-aminobenzaldehyde in 40 mL of absolute ethanol is heated to reflux with stirring. Copper(II) nitrate trihydrate (1.45 g, 0.006 mole) in 30 mL of absolute ethanol is added. The solution immediately changes from pale yellow to red-brown. The solution is refluxed for 1 hour, cooled to room temperature, and filtered. The dark-green microcrystalline product is washed with absolute alcohol and ether before drying *in vacuo* over P_4O_{10}. The yield is 2.9 g (81%). *Anal.* Calcd. for $C_{28}H_{20}N_6O_6Cu$: C, 56.04; H, 3.34; N, 14.01. Found: C, 56.51; H, 3.38; N, 13.95.

Properties

This dark-green crystalline complex is four-coordinate with a room-temperature moment of 1.84 BM.[4] Reduction of $[Cu(Bzo_4[16]octaeneN_4)]^{2+}$ with platinum oxide catalyst or with elemental mercury[18] leads to a deep-blue diamagnetic complex, $[Cu(Bzo_4[16]octaeneN_4)]^+$, which can also be obtained by electrochemical reduction. As is the case for the nickel complexes, $[Cu(Bzo_4[16]\text{-}octaeneN_4)]^{2+}$ is attacked by nucleophiles that add to two of the azomethine linkages of the macrocyclic ligand.[14]

D. (TETRABENZO[*b,f,j,n*][1,5,9,13]TETRAAZACYCLOHEXADECINE)-ZINC(II) TETRACHLOROZINCATE (2-)

$$2ZnCl_2 + 4(NH_2C_6H_4CHO) \longrightarrow [Zn(Bzo_4[16]octaeneN_4)][ZnCl_4] + 4H_2O$$

The self-condensation of *o*-aminobenzaldehyde in the presence of zinc chloride under anhydrous conditions was reported initially[12] to give a compound formulated as $C_7H_5N \cdot \frac{1}{2}ZnCl_2$. However, the compound has been shown to exhibit an infrared spectrum similar to those of the nickel and copper complexes of the tetrameric macrocyclic $Bzo_4[16]octaeneN_4$ so that this compound should have been formulated as $[Zn(Bzo_4[16]octaeneN_4)][ZnCl_4]$[4].

Procedure

A mixture of 42 g (0.308 mole) of anhydrous zinc chloride and 28.5 g (0.236 mole) of freshly prepared dry *o*-aminobenzaldehyde is stirred for 4 days in 250 mL of anhydrous diethyl ether* in a 500-mL stoppered Erlenmeyer flask equipped with a large magnetic stirring bar. The bright-yellow solid is removed from the diethyl ether by filtration and is air dried. This product is stirred for 30

*Strictly anhydrous materials and glassware are required.

minutes in 300 mL of water, again separated by filtration, washed with a few milliliters of water. This process is repeated with 300 mL of absolute ethanol. The isolated product is washed with a few milliliters of ethanol and dried *in vacuo* over P_4O_{10}. The yield id 21.4 g (53%). *Anal.* Calcd. for $C_{28}H_{20}N_4Cl_4Zn_2$: C, 49.08; H, 2.92; N, 8.18; Cl, 20.72. Found: C, 49.02; H, 2.73; N, 8.18; Cl, 19.35.

Properties

This square planar zinc complex is an excellent intermediate for the preparation of other Bzo_4 [16] octaene N_4 complexes by metal exchange, and its use alleviates the problem of separating the desired complex from other reaction side products when *o*-aminobenzaldehyde is condensed in the presence of the appropriate metal ion. Complexes of nickel, cobalt, iron, and palladium have been prepared from this zinc compound.[20]

E. DIBROMO(TETRABENZO[*b,f,j,n*][1,5,9,13]TETRAAZACYCLOHEXADECINE)COBALT(III) BROMIDE

$$CoBr_2 + 4(NH_2C_6H_4CHO) \xrightarrow{[O]} [Co(Bzo_4[16]octeneN_4)Br_2]Br + [Co(Bzo_3[12]hexeneN_3)_2]Br_3$$

As is the case for nickel(II),[4] the self-condensation of *o*-aminobenzaldehyde in the presence of cobalt(II) leads to the formation, on oxidation, of two types of macrocyclic Schiff-base cobalt(III) complexes containing either the ligand Bzo_4[16]octaeneN_4[9] or the tridentate macrocyclic ligand Bzo_3[12]hexaeneN_3.[10] Separation of the mixture into its components is easily accomplished by utilizing the marked solubility differences of the respective bromo complexes.[9]

Procedure

Freshly prepared *o*-aminobenzaldehyde (10.0 g, 0.083 mole) is dissolved in 40 mL of absolute ethanol. The solution is heated to reflux and 4.53 g (0.021 mole) of anhydrous cobalt(II) bromide in 10 mL of ethanol acidified with two drops of 48% hydrobromic acid is added. The solution immediately changes from yellow to green and solids begin to form after about 20 minutes. The solution is stirred and refluxed for 8 hours and then is chilled overnight. The dark-green product is collected, washed with cold ethanol and diethyl ether, and dried *in vacuo* for 24 hours at 100°. The yield is 5.7 g (44%). If air is bubbled through the reaction mixture during refluxing the product is a lighter green and the yield is increased by about 30%. The crude product (0.65 g) is stirred overnight at room temperature in 650 mL of ethanol, methanol, or water containing 65 mL of 48% hyrobromic acid. Air is bubbled through the the solution for the entire time and the red-brown precipitate that forms is collected by filtration. This product is a mixture of [Co(Bzo_4[16]octaeneN_4)-Br_2]Br and [Co(Bzo_3[12]hexaeneN_3)$_2$]Br_3. The red-brown product is washed with methanol containing 5% HBr to remove the soluble [Co(Bzo_3[12]octaene-N_3)$_2$]Br_3 complex until the filtrate is colorless. The maroon powder of [Co-(Bzo_4[16]octaeneN_4)Br_2]Br is highly insoluble and cannot be recrystallized. It is dried *in vacuo* for 24 hours at 100°. The yield is 0.14 g (19%). *Anal.* Calcd. for $C_{28}H_{20}N_4Br_3Co$: C, 47.27; H, 2.81; N, 7.87; Br, 33.73. Found: C, 46.93; H, 3.18; N, 7.83; Br, 33.87.

Properties

The self-condensation of *o*-aminobenzaldehyde in the presence of cobalt(II) leads to an initial dark-green crude product and a red-brown filtrate.[9] If air is bubbled through this filtrate, a brick-red powder can be isolated and has been characterized as the tetrabromocobaltate salt of the diprotonated metal-free tetramer, [$C_{28}H_{22}N_4$] [$CoBr_4$].[9] Oxidation of the initial crude green solid product leads to the maroon cobalt(III) complex of the tetramer, Bzo_4[16]-octaeneN_4 and to a yellow cobalt(III) complex of the type [Co(Bzo_3[12]hexa-eneN_3)$_2$]Br_3, where the ligand is the terdentate macrocycle.[10] The complex, [Co(Bzo_4[16]octaeneN_4)Br_2]Br, is six-coordinate and diamagnetic and behaves as a one-to-one electrolyte in dimethylformamide. The infrared spectrum shows bands at 1610, 1592, 1572 (C=N, imine), 1499, 1449, 795, and 765 cm^{-1} attributable to the Bzo_4[16]octaeneN_4 structure. The solid-state electronic spectrum shows poorly resolved bands 14.4, 18.9, 20.0, 23.6, and 27.0 kK, with solution spectral bands occurring at 15.2, 21.7, and 28.0 kK. The ligand field strength for the macrocycle has been calculated[9] as $Dq^{xy} = 2563$ cm^{-1}. Metathetical reactions[10] on the dibromo complex produce complexes of the type [Co(Bzo_4[16]octaeneN_4)X_2]X where X = Cl^-, NO_3^-, NCS^-, N_3^-, and NO_2^-.

References

1. S. G. McGeachin, *Can. J. Chem.*, **44**, 2323 (1966).
2. G. A. Melson and D. H. Busch, *Proc. Chem. Soc.*, **1963**, 223.
3. G. A. Melson and D. H. Busch, *J. Am. Chem. Soc.*, **86**, 4830 (1964).
4. G. A. Melson and D. H. Busch, *J. Am. Chem. Soc.*, **86**, 4834 (1964).
5. G. A. Melson and D. H. Busch, *J. Am. Chem. Soc.*, **87**, 1706 (1965).
6. L. T. Taylor, S. C. Vergez, and D. H. Busch, *J. Am. Chem. Soc.*, **88**, 3170 (1966).
7. L. T. Taylor and D. H. Busch, *J. Am. Chem. Soc.*, **89**, 5372 (1967).
8. L. T. Taylor and D. H. Busch, *Inorg. Chem.*, **8**, 1366 (1969).

9. S. C. Cummings and D. H. Busch, *J. Am. Chem. Soc.*, **92**, 1924 (1970).
10. S. C. Cummings and D. H. Busch, *Inorg. Chem.*, **10**, 1220 (1971).
11. I. Madden, Ph.D. Thesis, The Ohio State University, 1975.
12. F. Seidel, *Chem. Ber.*, **59B**, 1894 (1926).
13. L. T. Taylor, F. L. Urbach, and D. H. Busch, *J. Am. Chem. Soc.*, **91**, 1072 (1969).
14. V. Katović, L. T. Taylor, and D. H. Busch, *Inorg. Chem.*, **10**, 458 (1971).
15. L. I. Smith and J. W. Opie, *Org. Synth.*, **28**, 11 (1948).
16. S. W. Hawkinson and E. B. Fleischer, *Inorg. Chem.*, 8, 2402 (1969).
17. V. Katović, L. T. Taylor, F. L. Urbach, W. H. White, and D. H. Busch, *Inorg. Chem.*, **11**, 479 (1972).
18. N. E. Tokel, V. Katović, K. Farmery, L. B. Anderson, and D. H. Busch. *J. Am. Chem. Soc.*, **92**, 400 (1970).
19. N. E. Takvoryan, K. Farmery, V. Katović, E. S. Gore, F. V. Lovecchio, L. B. Anderson, and D. H. Busch, *J. Am. Chem. Soc.*, **96**, 731 (1974).
20. J. Skuratowicz, Ph.D. Thesis, The Ohio State University, 1973.

7. MACROCYCLIC TETRAAZATETRAENATO LIGANDS AND THEIR METAL COMPLEXES

Submitted by DENNIS P. RILEY* and DARYLE H. BUSCH*
Checked by DAVID E. FENTON† and RICHARD L. LINTVEDT†

There has been a continuing interest in the preparation of new synthetic macrocyclic ligands and in the systematic study of the chemistry of their metal complexes. The literature is replete with examples of current progress in this area,[1-4] and many reviews[5-11] covering various aspects of the synthesis of macrocyclic ligands and their metal complexes have been published. The detailed procedure of the synthesis of a 14-membered macrocyclic tetraazatetraene ligand, devoid of functional substituents, is described in full. The ligand is prepared by deacylation and demetalation of the dianionic ligand, first prepared by Jäger,[12-14] in the presence of nickel(II) cation (template reaction) to yield the neutral parent macrocyclic complex directly. The synthesis of the parent macrocyclic nickel complex is a four-step procedure involving the preparation of (1) 3-(ethoxymethylene)-2,4-pentanedione from common organic reagents, followed by (2) the condensation of either ethylenediamine or 1,3-propanediamine (trimethylenediamine) with 3-(ethoxymethylene)-2,4-pentanedione to give an organic material capable of acting as a dianionic linear quadridentate ligand, (3) formation of the neutral nickel(II) complex with the linear quadridentate ligand, and (4) reaction of the linear quadridentate nickel complex with additional diamine to afford the macrocyclic complex. Each of these steps is described in full. Also described in detail is the procedure used for the synthesis of the deacy-

*Department of Chemistry, The Ohio State University, Columbus, OH 43210.
†Department of Chemistry, Wayne State University, Detroit, MI 48202.

lated and demetalated ligand from the parent macrocyclic nickel complex and the procedure used to prepare a square planar nickel complex from the ligand so obtained.

A. 3-(ETHOXYMETHYLENE)-2,4-PENTANEDIONE[15]

$$HC(OCH_2CH_3)_3 + CH_3\overset{O}{\overset{\|}{C}}CH_2\overset{O}{\overset{\|}{C}}CH_3 + 2CH_3\overset{O}{\overset{\|}{C}}O\overset{O}{\overset{\|}{C}}CH_3 \longrightarrow$$

$$CH_3\overset{O}{\overset{\|}{C}}-\underset{\underset{CH(OCH_2CH_3)}{\|}}{C}-\overset{O}{\overset{\|}{C}}CH_3 + 2CH_3CO_2H + 2CH_3CO_2CH_2CH_3$$

Procedure

To a 1-L, one-necked, round-bottomed flask with a ground-glass joint are added 166 mL (1 mole) of triethyl orthoformate, 188 mL (2 moles) of acetic anhydride and 103 mL (1 mole) of 2,4-pentanedione. This mixture is heated rapidly to reflux (~135°) and is maintained there for 30 minutes under nitrogen, by which time a deep-red color forms. The solution is distilled under nitrogen until the temperature reaches 150°* (this initial distillate is discarded). The remaining solution is immediately and rapidly vacuum distilled (at 10 torr, the temperature of the distillate is 165-170°) and a pale-yellow liquid is collected. The distillation is stopped when the distillate begins to darken. (■ **Caution.** *The residue remaining in the reaction flask after the distillation is pyrophoric at elevated temperatures. The heating source should be removed, and nitrogen should be bled into the system. When the contents of the reaction flask have cooled to <100°, acetone or chloroform is added slowly to prevent the residue from solidfying.*) For best results in the succeeding step the yellowish liquid is used without further purification and immediately after collection. A typical yield is in the range 85-100 g (54-65% based on 2,4-pentanedione). The checkers obtained a yield of 70 g with unpurified reagents.

B. 3,3′-[ETHYLENEBIS(IMINOMETHYLIDYNE)]DI-2,4-PENTANEDIONE

$$2CH_3\overset{O}{\overset{\|}{C}}-\underset{\underset{CH(OCH_2H_3)}{\|}}{C}-\overset{O}{\overset{\|}{C}}CH_3 + NH_2CH_2CH_2NH_2 \longrightarrow$$

*Yields decrease significantly if the temperature rises above 150° even for short periods.

$$(CH_3\overset{O}{\overset{\|}{C}})_2C{=}CHNHCH_2CH_2NHCH{=}C(\overset{O}{\overset{\|}{C}}CH_3)_2 + 2H_2O$$

Procedure

One hundred grams (0.64 mole) of crude 3-(ethoxymethylene)-2,4-pentanedione (procedure A) is dissolved in 1 L of absolute ethanol in a 2-L Erlenmeyer flask. Absolute ethanol (500 mL) containing 19.2 g (0.32 mole) of freshly distilled (from KOH) anhydrous ethylenediamine is added dropwise over a 30-minute period. An immediate reaction is observed to occur with the formation of a white precipitate. The solution is then filtered and the fine white crystalline material is collected by filtration, washed with absolute ethanol and diethyl ether, and dried *in vacuo* over P_4O_{10} for several hours. The yield is 72 g (0.26 mole, 80%). The checkers obtained a 55% yield. This material is suitable for use without further purification. *Anal.* Calcd. for $C_{14}H_{20}N_2O_4$: C. 60.0; H, 7.15; N, 10.0. Found: C, 60.51; H, 7.21; N, 9.81.

C. {[3,3′-[ETHYLENEBIS(IMINOMETHYLIDYNE)]DI-2,4-PENTANEDIONATO](2−)}NICKEL(II)

$$(CH_3\overset{O}{\overset{\|}{C}})_2C{=}CHNHCH_2CH_2NHCH{=}C(\overset{O}{\overset{\|}{C}}CH_3)_2 + Ni(CH_3CO_2)_2{\cdot}4H_2O$$

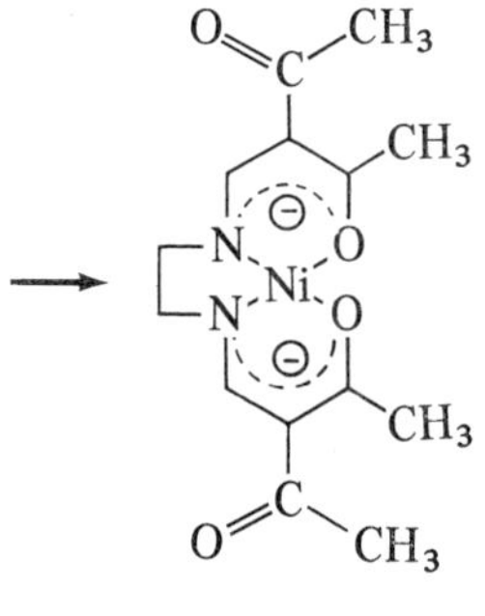

→ [nickel complex structure: O=C(CH₃), CH₃, N, O, Ni, N, O, ⊖, ⊖, CH₃, O=C(CH₃)] $+ 2CH_3CO_2H + 4H_2O$

Procedure

This square planar nickel complex is prepared by the reaction of the ligand prepared in the previous procedure (B above) with nickel(II) acetate tetrahydrate in warm (55°) methanol. In a typical preparation 74 g (0.265 mole) of the linear quadridentate ligand is slurried in 1 L of warm methanol in a 2-L flask. To this solution is added 65.0 g (0.26 mole) of nickel(II) acetate tetrahydrate with stirring. An orange precipitate forms immediately. The solution is stirred at 55° for 2 hours and is then filtered while still hot. The resulting orange, crystalline solid is washed with several 100-mL portions of hot methanol, followed by absolute diethyl ether. To remove all the solvent the orange needles are dried overnight *in vacuo* at 60°. When dry, this complex is suitable for use without further purification. The yield is 55 g (62%). *Anal.* Calcd. for $C_{14}H_{18}N_2NiO_4$: C, 49.89; H, 5.34; N, 8.32. Found: C, 49.61; H, 5.19; N, 8.39.

Properties

This square planar nickel(II) complex was first prepared by Jäger.[13] The complex is isolated as very stable orange needles. It is insoluble in water, ethanol, and methanol but soluble in less polar solvents, such as chloroform and dichloromethane. The complex shows two very broad and intense bands in its infrared spectrum, one centered at 1650 cm^{-1}, due to the C=O stretching mode, and the other centered at 1590 cm^{-1}, due to C=C and C=N stretching vibrations.

D. [6,13-DIACETYL-5,14-DIMETHYL-1,4,8,11-TETRAAZACYCLOTETRADECA-4,6,11,13-TETRAENATO(2−)]NICKEL(II) ([6,13-Ac_2-5,14-Me_2-[14]-4,6,11,13-tetraenato(2−)-1,4,8,11-N_4]Ni(II))

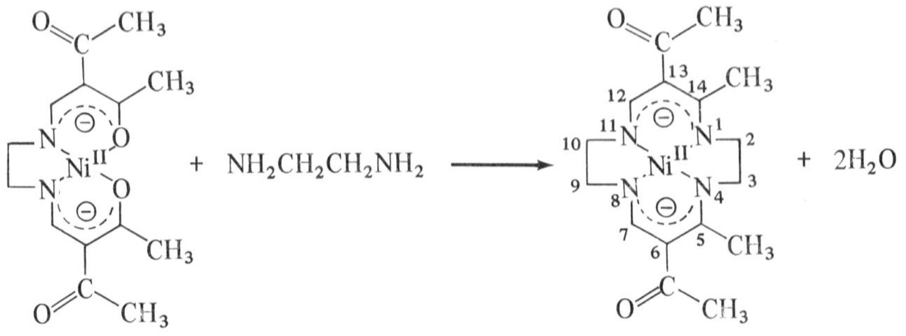

$$\text{[Ni(II) complex]} + NH_2CH_2CH_2NH_2 \longrightarrow \text{[Ni(II) macrocycle]} + 2H_2O$$

Procedure

In a typical preparation, 100 g (0.178 mole) of the linear quadridentate nickel(II) complex prepared in Section 7-C is placed in 1-L flask. To this is added 150 mL of dry ethylenediamine (dried by distillation over KOH). The resultant slurry is stirred and heated to reflux (~160°) under a blanket of nitrogen in an oil bath. After about ½ hour, all the solid dissolves to give a red solution, which is refluxed for an additional 15 minutes or until red crystals form. The solution is refluxed for a further 15 minutes and water (100 mL) is added after cooling. The red precipitate that forms is removed by filtration and is washed with copious amounts of water to remove traces of ethylenediamine. The red solid, after being air-dried overnight, is then dissolved in 500 mL of boiling 1:1 chloroform-methanol and the solution is filtered through Filter Aid to remove some nickel metal. The solvent is then removed by means of a rotary-flash evaporator and the product is washed into a filter funnel with methanol. It is washed with absolute diethyl ether and dried *in vacuo* at 60° overnight. To obtain satisfactory elemental analyses the product should be recrystallized three times and dried *in vacuo* at 60° for 24 hours. The yield after three recrystallizations is 30.6 g (0.085 mole), 48% based on starting complex. *Anal.* Calcd. for $C_{16}H_{22}N_4NiO_2$: C, 53.22; H, 6.10; N, 15.51. Found: C, 53.03; H, 6.00; N, 15.24.

Properties

This square planar nickel(II) complex is obtained as pink-red needles. The complex is slightly soluble in alcohols, insoluble in water, and very soluble in chloroform and dichloromethane. It is stable in air and is not hygroscopic. The compound shows a broad intense band centered at 1600 cm^{-1} in its infrared spectrum due to overlap of C=O and C=N stretching vibrations. The complex is diamagnetic and the PMR in $CDCl_3$ is fully consistent with the structure, showing four resonances at δ 7.5 (vinyl), δ 3.2 (methylene), δ 2.42 (methyl), and δ 2.26 (methyl).

E. 5,14-DIMETHYL-1,4,8,11-TETRAAZACYCLOTETRADECA-4,6,11,13-TETRAENE BIS(HEXAFLUOROPHOSPHATE) ([5,14-Me_2[14]-4,6,11,13-tetraene-1,4,8,11-N_4]·2HPF_6)*

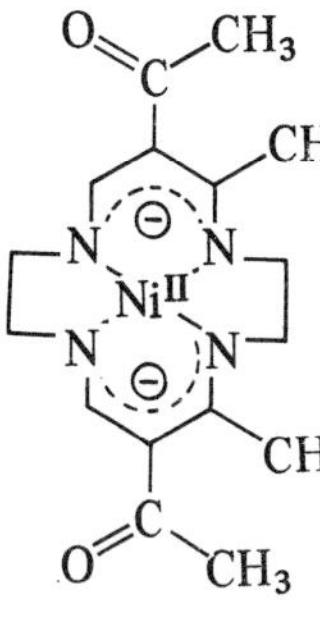

$$[\text{Ni complex}] + 2CH_3CH_2OH + 4HCl + 2NH_4PF_6 \longrightarrow$$

*For a discussion of the nomenclature of the macrocyclic ligands, and for the derivation of the abbreviations used to represent these ligands, see p. 2

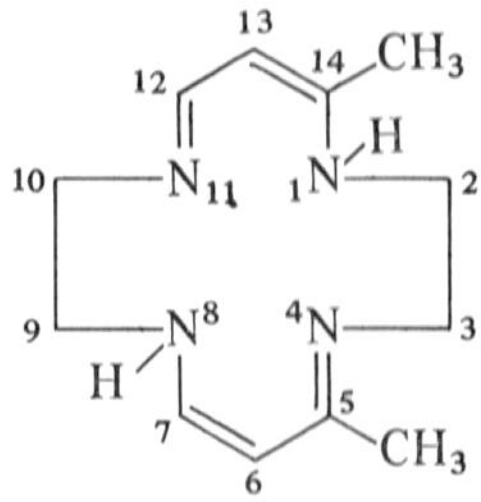

$\cdot 2HPF_6 + 2CH_3CO_2CH_2CH_3 + 2NH_4Cl + NiCl_2 \cdot aq$

*Procedure**

In a typical preparation of the ligand salt, 32.0 g (0.089 mole) of the macrocyclic nickel(II) complex prepared in Section 7-D is slurried in 200 mL of hot (80°) water. To this slurry 30-40 mL of 37% hydrochloric acid is added dropwise to dissolve the solids, and 32.2 g (0.198 mole) of $NH_4[PF_6]$ is added to the yellow solution. A yellow crystalline material forms and after an hour of cooling in an ice bath, the solution is filtered. Thirty-eight grams of the yellow crystalline bis(hexafluorophosphate) salt of the diprotonated macrocyclic nickel complex is obtained.† This complex is washed with absolute ethanol and diethyl ether and dried in a vacuum oven at 80° for 2 hours. The anhydrous crystalline salt is placed in a 1-L, round-bottomed flask and slurried with 500 mL of absolute ethanol. Hydrogen chloride gas is then bubbled through the solution for 2 hours. During this period the color changes from yellow to green to blue. The blue color, due to tetrahderal $[NiCl_4]^{2-}$ indicates that the reaction is completed. At this point all the ethanol is removed by using a rotary evaporator, and 100 mL of water is added to the blue solids. The blue tetrahedral tetrachloronickelate(2−) anion is destroyed in water and a yellow solution remains. To this solution is then added an additional 32.2 g of $NH_4[PF_6]$. A white precipitate forms immediately. After the flask is cooled in an ice bath for ½ hour, the precipitate is collected by filtration, and is washed first with water, several times with absolute ethanol, and then with diethyl ether. This cream- to tan-colored crystalline product is dried overnight *in vacuo* at 60°. The yield is 21.4 g (0.0425 mole). This corresponds to a 48% yield of ligand salt based on starting nickel complex.

*The method outlined in the preceding procedures for the preparation of the 14-membered macrocyclic ligand can be easily extended to the preparations of the corresponding 15- and 16-membered macrocyclic ligand salts. The 15-membered macrocyclic ligand can be prepared by substituting 1,3-propanediamine for ethylenediamine in the ring closure Sec. D. The 16-membered macrocyclic ligand can be prepared by substituting 1,3-propanediamine for ethylenediamine in both the preparation of the linear quadridentate ligand in Sec. B and the ring closure Sec. D.[4]

†The addition of two protons to the ligand in the nickel(II) complex is reversible[13] and occurs[4] on the apical carbon of the charged six-membered ring of the macrocycle.

The material is pure and no further recrystallizations are necessary. *Anal.* Calcd. for $C_{12}H_{18}N_4 \cdot 2HPF_6$: C, 28.13; H, 4.30; N, 10.94. Found: C, 28.19; H, 3.97; N, 10.87.

Properties

The cream-colored ligand salt is soluble in acetone, nitromethane, and acetonitrile but insoluble in water and alcohols. The ligand salt is a biunivalent electrolyte in acetonitrile and nitromethane. The infrared spectrum of this material shows several characteristic bands: a broad N–H stretching frequency at 3300 cm^{-1}, a broad intense absorption in the double-bond region at 1600 cm^{-1}, and bands due to hexafluorophosphate at 860 and 560 cm^{-1}. The PMR spectrum in nitromethane shows five types of proton resonances as expected: singlet methyl at δ 2.41, broad methylene at δ 3.83, doublet vinyl at δ 6.38 and δ 6.58, complex vinyl near δ 8.0, and broad NH at δ 7.6.

F. [5,14-DIMETHYL-1,4,8,11-TETRAAZACYCLOTETRADECA-4,6,11,13-TETRAENATO(2–)]NICKEL(II), [Ni(Me_2[14]-4,6,11,13-tetraenato-(2–)-N_4)]

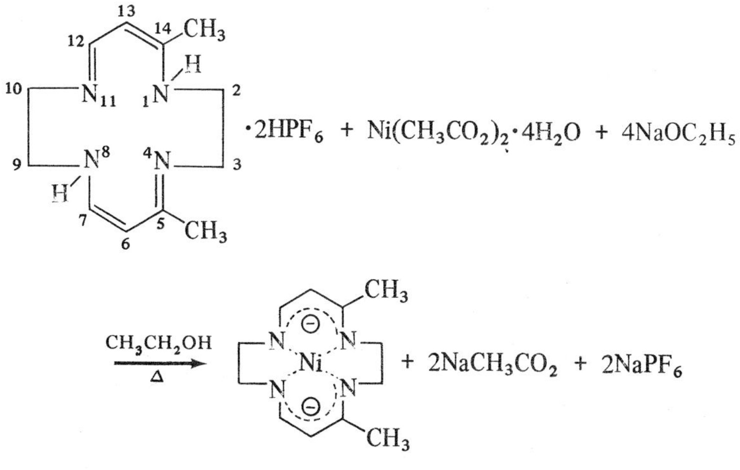

$\cdot 2HPF_6 + Ni(CH_3CO_2)_2 \cdot 4H_2O + 4NaOC_2H_5$

$\xrightarrow[\Delta]{CH_3CH_2OH}$ $+ 2NaCH_3CO_2 + 2NaPF_6$

*Procedure**

The preparation of this complex should be carried out in an inert atmosphere.

*The procedure for preparation of the square planar complexes of nickel(II) with the larger 15- and 16-membered macrocyclic ligands is identical to that outlined above for the 14-membered macrocyclic ring complex. In addition, square planar complexes of iron(II) and cobalt(II) have been prepared by the same procedure.[4]

In a typical preparation 10.0 g (0.0196 mole) of the macrocyclic ligand salt Me_2-[14]4,6,11,13-tetraene$N_4 \cdot 2HPF_6$ and 4.88 g(0.0196 mole) of nickel(II) acetate tetrahydrate are slurried with 100 mL of absolute ethanol. This mixture is heated to reflux and a freshly prepared ethanolic solution containing 0.080 mole of sodium ethoxide (prepared by adding 1.84 g of sodium to ethanol and allowing the sodium to react completely before using) is added to the slurry. An immediate reaction takes place and the color turns to a deep red. The solution is allowed to reflux for 1 hour, after which all solvent is removed under vacuum. One-hundred milliliters of benzene is added to the resulting dry solid. The contents of the flask are stirred with heating until reflux and the solution is filtered through a sintered-glass filter funnel containing a layer of Filter Aid to remove undesired contaminants. After filtration, the solvent is removed and 20 mL of boiling benzene is added to the solids. The resulting suspension is filtered through a sintered-glass filter funnel containing a 1-in. layer of Woelm neutral alumina. The volume of the resulting red benzene solution is reduced to about 2-3 mL and 20 mL of absolute ethanol is added. Red platelets of the complex crystallize upon cooling of the solution. These are collected by filtration and dried *in vacuo* for 2 hours. The yield is 2.1 g (7.6 mmole), 39% based on the ligand salt. *Anal.* Calcd. for $C_{12}H_{18}N_4Ni$: C, 52.04; H, 6.50; N, 20.24. Found: C, 51.88; H, 6.51; N, 20.12.

Properties

This square planar nickel(II) complex is stable in air in the solid state, but solutions of the complex react in air to produce oxidation products of unknown composition. For this reason, manipulations of the complex are best performed in an inert atmosphere. The complex is a nonelectrolyte and is soluble in most common organic solvents, including diethyl ether, but not in water. The infrared spectrum of the complex contains an intense broad band in the double-bond region centered at 1610 cm^{-1}. The electronic spectrum of a solution of the compound in toluene contains several bands: 17.9 (ϵ = 107), 23.0 ($\epsilon \sim 1600$), 24.5 ($\epsilon \sim 4600$), 25.9 ($\epsilon \sim 2700$), and 29.4 kK ($\epsilon \sim 5500$). The PMR spectrum of the complex in $CDCl_3$ contains four bands as expected: singlet methyl at δ 1.88, methylene at δ 3.13, vinyl doublet at δ 4.51, and a second vinyl doublet at δ 6.63. The vinyl protons of the charged chelate ring are coupled, J = 3 Hz.

References

1. S. C. Tang, S. Koch, G. N. Weinstein, R. W. Lane, and R. H. Holm, *Inorg. Chem.*, **12**, 2589 (1973).
2. L. Y. Martin, L. J. DeHayes, L. J. Zompa, and D. H. Busch, *J. Am. Chem. Soc.*, **96**, 4046 (1974).
3. F. V. Lovecchio, E. S. Gore, and D. H. Busch, *J. Am. Chem. Soc.*, **96**, 3109 (1974).

4. D. P. Riley, Ph.D. Thesis, The Ohio State University, 1975.
5. D. H. Busch, *Rec. Chem. Progr*, **25**, 107 (1964).
6. D. St. C. Black and E. Markham, *Rev. Pure Appl. Chem.*, **15**, 109 (1965).
7. A. B. P. Lever, *Adv. Inorg. Radiochem.*, **1**, 27 (1965).
8. D. H. Busch, *Helv. Chim. Acta* (fasciculus extraordinarius Alfred Werner), **1967**, 174.
9. N. F. Curtis, *Coord. Chem. Rev.*, **3**, 3 (1968).
10. L. F. Lindoy and D. H. Busch, *Preparative Inorganic Reactions*, Vol. 6, W. L. Jolly, (ed.), John Wiley and Sons, Inc., New York, 1971, p. 1.
11. D. H. Busch, K. Farmery, V. Goedken, V. Katović, A. C. Melnyk, C. R. Sperati, and N. Tokel, *Bioinorganic Chemistry*, Adv. Chem. Series, 100, American Chemical Society publication, 1971.
12. E. Jäger, *Z. Anorg. Allgem. Chem.*, **346**, 76 (1966).
13. E. Jäger, *Z. Chem.*, **8**, 30 (1968).
14. E. Jäger, *Z. Chem.*, **8**, 392 (1968).
15. L. Claisen, *Ann. Chem.*, **297**, 57 (1897).
16. J. C. Dabrowiak, P. H. Merrell, and D. H. Busch, *Inorg. Chem.*, **11**, 1979 (1972).
17. V. Goedken, P. H. Merrell, and D. H. Busch, *J. Am. Chem. Soc.*, **94**, 3397 (1972).

8. NONTEMPLATE SYNTHESES OF COMPLEXES WITH CONJUGATED MACROCYCLIC LIGANDS

Submitted by COLIN LUCAS HONEYBOURNE* and PAUL BURCHILL†
Checked by MYO K. YOO‡ and GEORGE R. BRUBAKER‡

Conjugated macrocyclic compounds play important roles in a number of important biological reactions[1] and there has been a resurgence of interest in the electrical properties of such compounds.[2] Although in both biochemical[1] and semisuperconducting applications,[3] side chains fulfill an important mechanistic role,[4] the central conjugated ring is clearly essential.[5,6] The compound 5,14-dihydrodibenzo[*b,i*] [1,4,8,11] tetraazacyclotetradecine (TADA-H_2) is a suitable parent system for biochemical and electrical studies. There are problems with some published syntheses and physical properties.[7]

*Bristol Polytechnic, Coldharbour Lane, Bristol, U.K.
†Science Research Council Post Doctoral Research Assistant.
‡Department of Chemistry, Illinois Institute of Technology, Chicago, IL 60616.

A. 5,14-DIHYDRODIBENZO[*b,i*][1,4,8,11]TETRAAZACYCLOTETRADECINE (2,3;9,10-Bzo_2[14]-2,4,6,9,11,13-hexaene-1,4,8,11-N_4 or TADA-H_2)

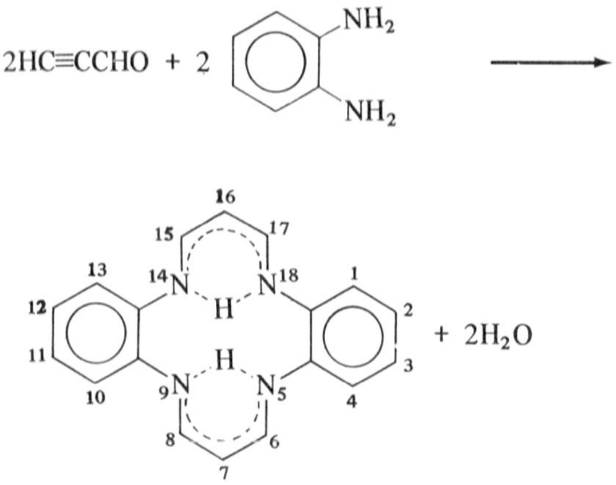

$$2HC{\equiv}CCHO + 2\ C_6H_4(NH_2)_2 \longrightarrow \text{TADA-}H_2 + 2H_2O$$

TADA-H_2

Numbering for abbreviation

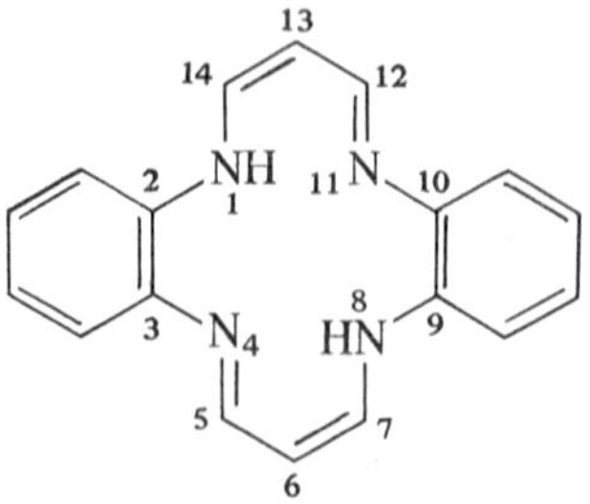

Procedure

Propynal (HC≡CCHO) (■ **Caution.** *Propynal is a lachrymator, with a foul odor. An efficient fume hood must be used.* o*-Phenylenediamine (1,2-benzenediamine) is toxic, and skin contact with the solid and inhalation either as vapor or fine dust must be avoided*) is prepared by standard methods[8] but is used without first resorting to fractional distillation. *o*-Phenylenediamine is crystallized from water after purification with activated charcoal and sodium dithionite at 100°; the diamine is sublimed before use. We strongly emphasize the care needed in diluting the propynal and that needed in adding a solution of the diamine.

(■ **Caution.** *The enthalpy of mixing is high and the reaction is notably exothermic).*

A solution of 25.50 g (0.236 mole) of *o*-phenylenediamine in 30 mL of dimethylformamide is placed in a three-necked reaction flask fitted with a Liebig condenser, a dropping funnel with a fine drawn-out dropping tube, and a thermometer. The flask is cooled in an ice-salt-water bath. Propynal, 12.71 g (0.235 mole), is diluted cautiously with 10 mL of methanol and the solution is added very slowly to the vigorously stirred diamine solution. The temperature tends to rise very rapidly at first but is kept below 10°. After the addition is complete (1-2 hr) the dark-red mixture is heated slowly to boiling (96°) and refluxed for 2 hours. The mixture is cooled, filtered, washed with a large quantity of cold methanol, and dried by suction. The deep-purple crystalline powder (brown on crushing) is dried further *in vacuo* over silica gel. The yield is 6.44 g (0.022 mole; 19.0%) *Anal.* Calcd. for $C_{18}H_{16}N_4$: C, 74.96; H, 5.59; N, 19.43. Found: C, 74.45; H, 5.68; N, 19.40.

Properties

The electronic spectrum of the product is as follows: 433 nm, ϵ_M = 19.400; 417 nm, ϵ_M = 21,800; 364 nm, ϵ_M = 57,300 nm; 352 nm, ϵ_M = 42,500; 262 nm, ϵ_M = 25,600; mp 296-299° (lit). The molar diamagnetic susceptibility of -140×10^{-6} cgs emu is smaller than expected, which suggests that there is a Van Vleck contribution to the induced magnetic moment.[9]

B. [5,14-DIHYDRODIBENZO[*b,i*][1,4,8,11]TETRAAZACYCLOTETRADECINATO(2−)]COBALT(II) [Co(II)TADA]

$$\text{TADA-H}_2 + \text{M(CH}_3\text{COO)}_2 \longrightarrow \text{M-TADA} + 2\ \text{CH}_3\text{COOH}$$

(M = Ni, Co, Cu)

Procedure

Cobalt(II) acetate tetrahydrate (1.89 g, 0.0075 mole) is dissolved in 30 mL of warm dimethylformamide, giving a purple solution. This is added to a stirred slurry of 1.81 g (0.0063 mole) of TADA-H_2 in 20 mL of hot dimethylformamide. The resulting red-brown mixture is heated under gentle reflux for 30 minutes and cooled to room temperature before being placed in an ice-water bath. Filtration yields a light-red residue, which is retained, and a red-black fitrate, which is discarded. After washing with a large quantity of cold methanol, the solid complex is dried *in vacuo* over silica gel. Yield 1.91 g (0.0055 mole,

87%; yield based on propynal 16.5%). *Anal.* Calcd. for $C_{18}H_{14}N_4Co$: C, 62.61; H, 4.09; N, 16.23. Found: C, 62.86; H, 4.08; N, 16.54.

Properties

The complex may be recrystallized from boiling dimethylformamide; the very small crystals have a metallic, slightly purple, sheen. On grinding, a red-brown powder is obtained. The complex is paramagnetic, in contrast to the observations by Hiller et al.[7] The very rapid electron spin relaxation in low-spin cobalt(II) systems prevents the detection of paramagnetism by EPR at room temperature. We find μ_{eff} = 2.44 BM at 297°[5].

C. 5,26:13,18-DIIMINO-7,11:20,24-DINITRILODIBENZO[*c,n*][1,6,12,17]-TETRAAZACYCLODOCOSINE [Hp-H_2][10,11]

$$2C_6H_4(CN)_2 + 2C_5H_3N(NH_2)_2 \longrightarrow 2NH_3 +$$

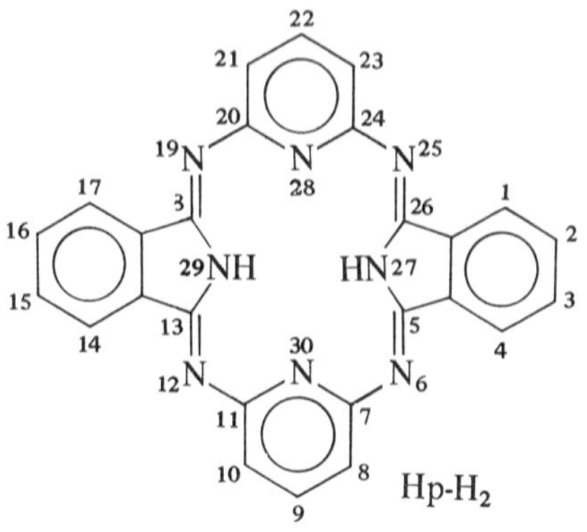

Hp-H_2

Procedure

A slurry of 16.0 g of *o*-phthalonitrile (1,2-benzenedicarbonitrile) (0.125 mole) (■ **Caution.** *Toxic*) and 13.63 g of 2,6-diaminopyridine (0.125 mole) in 100 mL of 1-chloronaphthalene is heated under gentle reflux for 24 hours. The tarry reaction mixture is allowed to cool and is diluted with 100 mL of cold methanol and shaken vigorously for 1 hour. The required solid is filtered using a sintered-glass filter and washed with cold methanol until the washings are pale yellow. The metallic-looking, copper-red needles are dried by suction and stored *in* *vacuo* over silica gel. The yield is 14.85 g (0.0337 mole, 54%). *Anal.* Calcd. for $C_{26}H_{16}N_8$: C, 70.90; H, 3.36; N, 25.45. Found: C, 70.67; H, 3.89; N, 25.56; mp 344° (lit). If required, the macrocycle may be recrystallized from boiling nitrobenzene (4 g of Hp-H_2 with 150 mL of solvent) (■ **Caution.** *Fume hood*). The nitrobenzene is solvated tenaciously to the macrocycle.

D. [5,26:13,18-DIIMINO-7,11:20,24-DINITRILODIBENZO[*c,n*][1,6,12,17]-TETRAAZACYCLODOCOSINATO(2−)]MANGANESE(II) [Hp-Mn(II)]

$$\text{Hp-H}_2 + \text{M(CH}_3\text{COO)}_2 \longrightarrow [\text{M-Hp}] + 2\text{CH}_3\text{COOH}$$

(M = Mn, Co, Ni, Cu)

Procedure

Manganese diacetate tetrahydrate (2.69 g, 0.011 mole) is dissolved in 30 mL of dimethylformamide. This solution is added to a slurry of 4.03 g (0.0091 mole) of Hp-H_2 in 30 mL of hot dimethylformamide. The dark-green mixture is heated under gentle reflux for 3½ hours. The reaction mixture is cooled and filtered and solid product is washed with cold dimethylformamide and then with cold methanol. The microcrystalline dark-blue powder is dried by suction and then stored *in vacuo* over silica gel. Yield of the monohydrate, 3.77 g (0.0074 mole, 81%). *Anal.* Calcd. for $C_{26}H_{16}N_8OMn$: C, 61.08; H, 3.15; N, 21.93. Found: C, 61.4; H, 3.43; N, 21.90.

Properties

The extremely insoluble, oxygen- and moisture-insensitive complex obeys the Curie law between 85 and 300°K, with a constant effective magnetic moment of 5.80 BM. The magnetic properties of the nickel, cobalt, and copper analogues are given elsewhere.[5]

E. [5,26:13,18-DIIMINO-7,11:20,24-DINITRILODIBENZO[*c,n*][1,6,12,17]-TETRAAZACYCLODOCOSINATO(2−)]OXOVANADIUM(IV)

$$\text{Hp-H}_2 + \text{(VO)SO}_4 \longrightarrow [\text{VO-Hp}] + \text{H}_2\text{SO}_4$$

Procedure

Oxovanadium(IV) sulfate pentahydrate (vanadyl(2+) sulfate pentahydrate) (2.79 g, 0.011 mole) is dissolved in 30 mL of boiling dimethylformamide (■ **Caution.**

Fume hood), and the dark-blue solution so formed is added to a slurry of 4.01 g (0.0091 mole) of Hp-H_2 in 30 mL of hot dimethylformamide. The grey-green mixture is refluxed gently, with stirring, for 4 hours and allowed to cool. After filtration the solid residue is washed thoroughly with cold dimethylformamide and then with cold methanol. The olive-green powder is dried by suction and stored *in vacuo* over silica gel. Yield of the dihydrate is 2.53 g (0.0047 mole, 51%). *Anal.* Calcd. for $C_{26}H_{18}N_8O_3V$: C, 57.69; H, 3.35; N, 20.70. Found: C, 57.76; H, 2.95; N, 21.06.

Properties

The very insoluble, oxygen- and moisture-insensitive complex obeys the Curie law, after allowance is made for TIP corrections,[12] with an effective magnetic moment of 1.87 BM at 300°K.

References

1. J. E. Falk, *Porphyrins and Metalloporphyrins*, Elsevier, London, 1964.
2. J. H. Hodgkin and J. Heller, *J. Polymer. Sci. C*, **29**, 37 (1970).
3. W. A. Little, *J. Polymer Sci. C*, **29**, 17 (1970).
4. M. F. Perutz, *Nature*, **237**, 495 (1972).
5. C. L. Honeybourne and P. Burchill, *Inorg. Nucl. Chem. Lett.*, **10**, 715 (1974).
6. C. L. Honeybourne, *Tetrahedron*, **29**, 1549 (1973).
7. H. Hiller, P. Dimroth and H. Pfitzner, *Ann. Chem.*, **717**, 137 (1968).
8. J. C. Sauer, *Org. Synth.*, **36**, 67 (1956).
9. C. L. Honeybourne, *Tetrahedron Lett.*, **1974**, 3075.
10. J. A. Elvidge and R. P. Linstead, *J. Chem. Soc.*, **1952**, 5008.
11. J. N. Esposito, L. E. Sutton, and M. E. Kenney, *Inorg. Chem.*, **6**, 1116 (1967).
12. B. N. Figgis and J. Lewis, *Progr. Inorg. Chem.*, **6**, 37 (1964).

9. TEMPLATE SYNTHESES OF COMPLEXES WITH PARTIALLY UNSATURATED MACROCYCLIC LIGANDS

Submitted by COLIN LUCAS HONEYBOURNE*
Checked by MICHAEL GERY† and BRADFORD B. WAYLAND†

There are macrocyclic ligands of biochemical interest that contain a modified porphine skeleton in varying degrees of reduction compared to the porphine nucleus; one such species is tetrahydrocorrin.[1,2] The reaction between (1,2-

*Bristol Polytechnic, Coldharbour Lane, Bristol, U.K.
†Department of Chemistry, University of Pennsylvania, Philadelphia, PA 19174.

alkanediamine)(1,2-arylenediamine)metal(II) diacetates or dichlorides and 2-malonaldehydes with electron-withdrawing substituents has been successful only when the complexes of copper(II) diacetate[3] are used.

A. BROMOMALONALDEHYDE (Bromopropanedial)

$$Br_2 + (C_2H_5O)_2CHCH_2CH(C_2H_5O)_2 + 2H_2O \xrightarrow{\text{acid}}$$
$$OHCCHBrCHO + 4C_2H_5OH + HBr$$

Procedure

Concentrated hydrochloric acid (20 mL) is added to a vigorously stirred solution of 100 mL (0.34 mole) of 1,1,3,3-tetraethoxypropane [malonaldehyde bis-(diethyl acetal)] in 100 mL of distilled water. To this is added, dropwise, 17.5 mL (0.34 mole) of bromine. (■ **Caution.** *Use a fume hood.*) The coloration of the bromine disappears immediately if the dropping rate is sufficiently slow. No increase in temperature is observed.

The reaction mixture is taken to a slush in a rotary evaporator (55°/25 torr) and filtered through a sintered-glass filter. The pale-yellow, crystalline product is dried by suction and stored *in vacuo* over silica gel. Yield 29.35 g (0.19 mole, 57%) *Anal.* Calcd. for $C_3H_3O_2Br$: C, 23.87; H, 2.00; Br, 52.93. Found: C, 23.95; H, 1.88; Br, 53.2.

B. [3,10-DIBROMO-1,6,7,12-TETRAHYDRO-1,5,8,12-BENZOTETRAAZA-CYCLOTETRADECINATO(2−)]COPPER(II) (Cu(II)[6,13-Br$_2$-2,3-Bzo[14]-2,4,6,11,13-pentaenato(2−)-1,4,8,11-N$_4$])

$$[(CH_3COO)_2Cu(en)C_6H_4(NH_2)_2] + 2BrCH(CHO)_2 \longrightarrow$$

$$2CH_3COOH + 4H_2O +$$

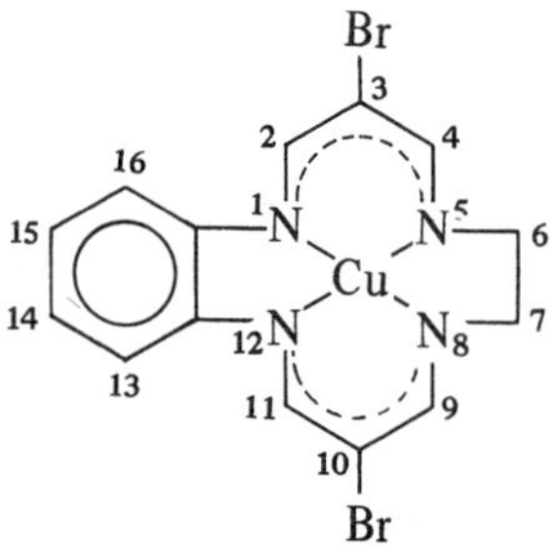

en = $H_2NC_2H_4NH_2$

Numbering for abbreviation

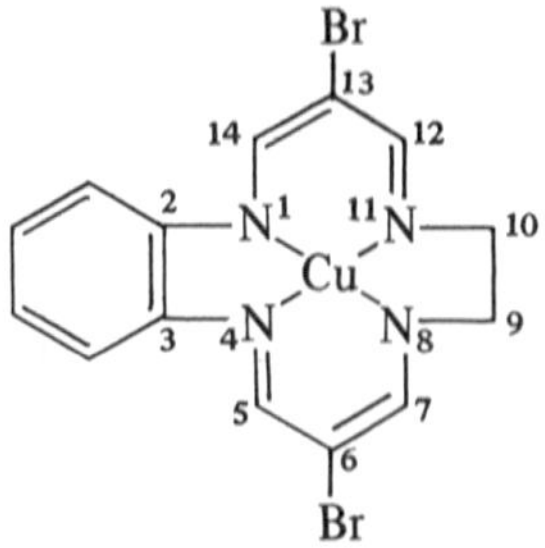

Procedure

A solution of 1.08 g (0.01 mole) of *o*-phenylenediamine (1,2-benzenediamine) in 100 mL of dry ethanol is added very slowly with vigorous stirring to a solution of 2.0 g (0.01 mole) of reagent grade copper(II) acetate. The dark-green precipitate is filtered through a sintered-glass filter *without* suction, washed with dry ethanol *without* suction, and then washed into a 250-mL conical flask with approximately 100 mL of dry ethanol from a wash bottle. To this suspension of the monoamine complex, a solution of 0.65 mL (0.01 mole) of ethylenediamine (1,2-ethanediamine) in 50 mL of dry ethanol is added dropwise over an extended period at 10°. The mixture is allowed to remain in the dark for 24 hours at this temperature, with constant vigorous stirring.

The reaction mixture is taken to 2° and a solution of 3.08 g (0.02 mole) of bromomalonaldehyde in 50 mL of dry ethanol is added. This green reaction mixture is stirred vigorously in the dark for 7 days at 2° until the suspended solid is black. This solid is filtered, washed thoroughly with dry ethanol, and dried *in vacuo* over silica gel. The yield of anhydrous product is 2.13 g (0.0046 mole, 46%). *Anal.* Calcd. for $C_{14}H_{12}N_4CuBr_2$: C, 36.59; H, 2.63; N, 12.19. Found: C, 36.60; H, 2.53; N, 12.57.

Properties

The very insoluble black solid affords the required peaks in the mass spectrum for $C_{14}H_{12}N_4CuBr_2$ at 280° with a beam strength of 19 eV. The base peak is given by $H^{79}Br$.

General Remarks

The low reaction temperature is used to prevent the reaction between free *o*-phenylenediamine and the dialdehyde, which gives the macrocycle (7,16-dibromo-5,14-dihydrodibenzo[*b,i*] [1,4,8,11] tetraazacyclotetradecine) even in the absence of metal ions.[4] The above reaction scheme is also successful when $NO_2CH(CHO)_2$ or $C_2H_5OOCCH(CHO)_2$ is used.

References

1. R. L. N. Harris, A. W. Johnson, and I. T. Kay, *Chem. Commun.*, **1965**, 355.
2. E. G. Jaeger, *Z. Chem.*, 8, 30 (1968).
3. C. L. Honeybourne, *Inorg. Nucl. Chem. Lett.*, **11**, 191 (1975).
4. C. L. Honeybourne, *Chem. Ind. (London)*, **1975**, 350.

Chapter Two

METAL CARBONYL COMPLEXES

10. MOLYBDENUM(II) CARBONYL COMPLEXES CONTAINING THIO LIGANDS AND ACETYLENE

Submitted by W. E. NEWTON*, JAMES L. CORBIN*, and JOHN W. MCDONALD*
Checked by FREDERICK I. KEEN† and T. M. BROWN†

Organometallic complexes of molybdenum containing sulfur donor ligands but without π-bonded entities, such as η-cyclopentadienyl and η-benzene, have recently become of interest as probes for the interactions occurring at the metal-containing sites of various molybdoenzymes.[1,2] The following procedure may be used to prepare a range of such complexes involving acetylene or substituted acetylenes.[3]

A. DICARBONYLBIS(DIISOPROPYLPHOSPHINODITHIOATO)MOLYBDENUM(II)

$$Mo(CO)_6 + Cl_2 \longrightarrow Mo(CO)_4Cl_2 + 2CO$$

$$Mo(CO)_4Cl_2 + (i\text{-}C_3H_7)_2P(S)SH \longrightarrow Mo(CO)_2[S_2P(i\text{-}C_3H_7)_2]_2 + 2CO + 2HCl$$

*Charles F. Kettering Research Laboratory, Yellow Springs, OH 45387 Contribution No. 577.
†Department of Chemistry, Arizona State University, Tempe, AZ 85281.

Procedure

■ **Caution.** *The toxicity of chlorine makes the use of a fume hood mandatory.*

An argon-filled, 200-mL Schlenk tube containing solid $Mo(CO)_6$ (Pressure Chemical Co.) (4.0 g, 23.1 mmole) is fitted with a rubber serum stopper and immersed in a Dry Ice-acetone bath. While the system is still under positive argon pressure, chlorine inlet and venting tubes are inserted into the serum stopper and Cl_2 (about 30 mL of liquid) is condensed onto the solid. After condensation is complete, the yellow slurry is magnetically stirred at -78° for 30 minutes, the bath is removed, and the Cl_2 is allowed to evaporate at room temperature in a stream of argon. The resulting solid residue* is subjected to pumping for 5 minutes to remove any residual Cl_2 and then dissolved in methanol (70 mL) that has been deaerated by purging with argon. (■ **Caution.** *$Mo(CO)_4Cl_2$ dissolves with very vigorous gas evolution. The reaction vessel must be vented during this step. The complex $Mo(CO)_4Cl_2$ is also thermally unstable, decomposing completely in a few hours at room temperature to a dark powder that is sometimes pyrophoric when exposed to air. For this reason the compound should be dissolved and the ligand added within a few minutes after removal of Cl_2.*) A deaerated solution of diisopropylphosphinodithioic acid ($((i\text{-}C_3H_7)_2P(S)(SH))$) (6.5 mL, about 35.7 mmole)[4] in methanol (75 mL) is added to the above solution giving a deep-red reaction mixture containing $Mo(CO)_3[S_2P(i\text{-}C_3H_7)_2]_2$. The solution is filtered and the filtrate is evaporated under vacuum to a volume of about 45 mL. During the evaporation the solution turns green and the product separates as a green crystalline solid, which is isolated by filtration, washed with deaerated methanol (3 × 30 mL), and dried *in vacuo*. The yield is 6.0 g, 77% based on $Mo(CO)_6$. *Anal.* Calcd. for $C_{14}H_{28}O_2P_2S_4Mo$: C, 32.7; H, 5.45; CO, 10.9. Found: C, 32.3; H, 5.45; CO, 10.6.

Properties

The product is a green crystalline solid and is air sensitive in solution or in the solid state. Its infrared spectrum contains strong bands at 1860 and 1960 cm^{-1} that are assigned to carbonyl stretching frequencies. In most preparations the infrared spectrum of the final product also contains a weak band at 2020 cm^{-1}, which may be due to slight contamination with $Mo(CO)_3[S_2P(i\text{-}C_3H_7)_2]_2$.

The compound $Mo(CO)_2[S_2P(i\text{-}C_3H_7)_2]_2$ is monomeric (cryoscopy) and thus is a coordinatively unsaturated, 16-electron species that is potentially very reactive. The compound reacts reversibly with CO and triphenylphosphine to

*The residue is predominantly yellow but may contain small amounts of green or blue impurities arising from thermal decomposition of $Mo(CO)_4Cl_2$. These impurities are of little consequence and are removed in the subsequent filtration.

form $Mo(CO)_3[S_2P(i\text{-}C_3H_7)_2]_2$ and $Mo(CO)_2[P(C_6H_5)_3][S_2P(i\text{-}C_3H_7)_2]_2$, respectively, with diazenes and hydrazines, and with a variety of acetylenes (see Sec. 10-B).

By modifying the above procedure slightly, the synthesis can be used to prepare complexes containing *N,N*-dialkyldithiocarbamato ligands.[5] Such complexes were reported as intermediates in the synthesis[6] of $Mo(NO)_2(S_2CNR_2)_2$ but were not isolated. The $Mo(CO)_2(S_2CNR_2)_2$ complexes show similar reactivity[7] to that described above for $Mo(CO)_2[S_2P(i\text{-}C_3H_7)_2]_2$.

B. (ACETYLENE)CARBONYLBIS(DIISOPROPYLPHOSPHINODITHIOATO)MOLYBDENUM(II)

$$Mo(CO)_2[S_2P(i\text{-}C_3H_7)_2]_2 + C_2H_2 \longrightarrow$$
$$Mo(CO)(C_2H_2)[S_2P(i\text{-}C_3H_7)_2]_2 + CO$$

Procedure

Acetylene is bubbled for 30 minutes into a solution of $Mo(CO)_2[S_2P(i\text{-}C_3H_7)_2]_2$ (2.05 g, 3.99 mmole) in deaerated CH_2Cl_2 (50 mL) under argon. A Schlenk tube fitted with a rubber serum stopper (through which acetylene inlet and venting needles are inserted) is used as the reaction vessel. The reaction mixture initially turns brown and then slowly becomes dichroic,* finally appearing green in reflected and red in transmitted light. After filtration to remove a small amount of solid, the reaction mixture is evaporated to dryness under vacuum, yielding a yellow-green tacky residue. Trituration with deaerated hexane (75 mL) gives a red supernatant liquid and a yellow-green solid which is isolated by filtration, washed with deaerated hexane (3 × 30 mL), and dried *in vacuo*. The yield is 1.80 g, 88% based on $Mo(CO)_2[S_2P(i\text{-}C_3H_7)_2]_2$. *Anal.* Calcd. for $C_{15}H_{30}OP_2S_4Mo$: C, 35.2; H, 5.86. Found: C, 34.8; H, 6.24.

Properties

The compound $Mo(CO)(C_2H_2)[S_2P(i\text{-}C_3H_7)_2]_2$ is a yellow-green solid and is air sensitive in solution or in the solid state. Its infrared spectrum (KBr) contains bands at 1960 (ν_{CO}) and 3070, 3150, and 745 cm^{-1} (assigned to the C—H of the coordinated acetylene). No carbon-carbon stretch is observed in the usual region, suggesting a significant perturbation of the triple bond. The NMR spectrum

*The degree of dichroism varies with the preparation. The phenomenon appears to be due to small amounts of a red impurity (either in $Mo(CO)_2[S_2P(i\text{-}C_3H_7)_2]_2$ or generated in the reaction), which is removed in the hexane extraction.

($CDCl_3$) of $Mo(CO)(C_2H_2)[S_2P(i\text{-}C_3H_7)_2]_2$ exhibits in addition to a complex signal for the isopropyl groups, a sharp singlet *12.33 ppm downfield* from tetramethylsilane that is assigned to the two protons of the coordinated acetylene. These unusual spectral properties can be rationalized by treating acetylene as a 4-electron donor to the 14-electron $Mo(CO)[S_2P(i\text{-}C_3H_7)_2]_2$ core.

A series of compounds of the form $Mo(CO)(R{-}C_2R')[S_2P(i\text{-}C_3H_7)_2]_2$ (RC_2R' = $CH_3C{\equiv}CH$, $C_6H_5C{\equiv}CH$, $CH_3C{\equiv}CC_6H_5$, $C_6H_5C{\equiv}CC_6H_5$, $CH_3O_2CC{\equiv}CCO_2CH_3$, $HC{\equiv}CCO_2CH_3$) can be synthesized similarly using a 10% molar excess of the solid or liquid acetylene. The spectral data for these complexes are very similar to those described for $Mo(CO)(C_2H_2)[S_2P(i\text{-}C_3H_7)_2]_2$.

References

1. W. E. Newton and C. J. Nyman (eds.), *Proceedings of the International Symposium on Nitrogen Fixation,* Washington State University Press, Pullman, Washington, 1976, pp. 53-74.
2. E. I. Stiefel, *Proc. Natl. Acad. Sci. U.S.*, 70, 988 (1973).
3. J. W. McDonald, J. L. Corbin, and W. E. Newton, *J. Am. Chem. Soc.*, 97, 1970 (1975).
4. J. L. Corbin, W. E. Newton, and J. W. McDonald, *Organ. Prep. Proc. Int.*, 7, 309 (1975).
5. R. Colton, G. R. Scollory, and I. B. Tomkins, *Aust. J. Chem.*, **21**, 15 (1968).
6. J. A. Broomhead and W. Grumley, *Inorg. Synth.*, **16**, 235 (1976).
7. J. W. McDonald, W. E. Newton, C. T. C. Creedy, and J. L. Corbin, *J. Organometal. Chem.*, **92**, C25(1975).

11. TRIS[*cis*-(DIACETYLTETRACARBONYLMANGANESE)]-ALUMINUM [Hexa-μ-acetyl-tris(tetracarbonylmanganese)-aluminum]

Submitted by C. M. LUKEHART,* G. PAULL TORRENCE,* and JANE V. ZEILE*
Checked by B. DUANE DOMBEK† and ROBERT J. ANGELICI†

Tris[*cis*-(diacetyltetracarbonylmanganese)]aluminum is prepared readily by treating acetylpentacarbonylmanganese with 1 molar-equivalent of methyllithium at 0° followed by the addition of ⅓ molar-equivalent of anhydrous aluminum chloride. This complex is isostructural with tris(2,4-pentanedionato)aluminum (where 2,4-pentanedione ≡ acetylacetone) except that the methine group is replaced formally by a $Mn(CO)_4$ group, which suggests that the title compound is one example of a "metallo-β-diketonate" type complex.

*Department of Chemistry, Vanderbilt University, Nashville, TN 37235.
†Department of Chemistry, Iowa State University, Ames, IA 50010.

This structural similarity has been confirmed by X-ray crystallography.[1,2] The metallo-β-diketonate complex possesses rigorous crystallographic C_2 symmetry and idealized D_3 molecular symmetry. The central aluminum atom has only slightly distorted octahedral coordination and the metallo-β-diketonate ligand is symmetrical and essentially planar. The oxygen-oxygen chelating bite distance of 2.74 Å of the metallo-β-diketonate ligand is identical to the corresponding distance found in $Al(acac)_3$ (acac = acetylacetonato).

This procedure has been shown recently to be a general preparative method for the synthesis of a large variety of metallo-β-diketonate complexes of aluminum.[3] In each case a diacylmetalate anion is prepared from an acyl complex and is then complexed to the aluminum atom. The preparation of the metallo-β-diketonate complex presented here utilizes acetylpentacarbonylmanganese as the acyl complex. The preparation of this acetyl complex from acetyl chloride and sodium pentacarbonylmanganate(1−) is provided, also.

A. ACETYLPENTACARBONYLMANGANESE

$$Mn_2(CO)_{10} + 2Na/Hg \longrightarrow 2Na[Mn(CO)_5]$$

$$Na[Mn(CO)_5] + CH_3C(O)Cl \longrightarrow [[CH_3C(O)]Mn(CO)_5] + NaCl$$

Procedure

A 100-mL two-necked flask is removed from a 130° drying oven and flushed well with prepurified nitrogen, after a gas inlet is attached to the side neck. This flask is charged with 4 mL of mercury and a stirring bar, and then 0.50 g (21.7 mmole) of sodium metal is added to the stirred mercury puddle in small pieces, one at a time, under a continuous nitrogen flush. (■ **Caution.** *This operation should be done in the hood since the dissolution of the sodium metal in mercury is a highly exothermic reaction.*) After the sodium amalgam has cooled to room temperature, 50 mL of freshly distilled tetrahydrofuran[4] is introduced, followed by the addition of 3.0 g (7.7 mmole) of decacarbonyldimanganese.* (■ **Caution.** *Metal carbonyl compounds are extremely toxic and should be handled in an efficient hood.*) The yellow solution is stirred at 25° under nitrogen for 75 minutes.

After this time the solution of $Na[Mn(CO)_5]$ is transferred by means of a syringe into another 100-mL, two-necked flask that contains a stirring bar and has been flushed well with nitrogen. The solution is cooled to −78° (Dry Ice-

*Decacarbonyl dimanganese was purchased from Pressure Chemical Co., Pittsburgh, PA 15201.

acetone bath), and 1.2 mL (17.0 mmole) of acetyl chloride is added with a syringe from a freshly opened bottle. The reaction solution is stirred at −78° for 1 hour, the bath is removed, and the stirring is continued for 1 hour more. The solvent is removed at reduced pressure (5 torr, 25°), and the solid residue is stirred with 100 mL of hexane for 30 minutes at 25°. The hexane solution is filtered through a Schlenk frit,[5] and the filtrate is cooled at −20° for 16 hours. The crystallized solid is collected on a glass frit in air and dried briefly, using a water aspirator, giving 1.75 g (48%) of an off-white solid. A second crop of slightly impure product (about 0.5 g) is obtained similarly by concentrating the hexane filtrate to one-half the original volume, followed by cooling at −78° for 4 hours. *Anal.* Calcd. for $C_7H_3MnO_6$: C, 35.29; H, 1.26. Found: C, 35.10; H, 1.35.

Properties

Acetylpentacarbonylmanganese is a moderately volatile, white solid, m.p. 54.5-56°. It is air stable over at least a 2-day period. It has excellent solubility in most organic solvents. The infrared spectrum in cyclohexane* shows ν(C≡O) bands at 2114 (w), 2049 (w), 2011 (vs), 2002 (s) and a ν(acyl) band at 1663 (s) cm^{-1}. The ^1H NMR spectrum ($CDCl_3$ versus TMS) shows a singlet for the methyl resonance at τ7.43. The dipole moment, measured in benzene solution, is 2.27 (5) D.[6] Acetylpentacarbonylmanganese has been used as a catalyst for a hydroformylation reaction.[7]

B. TRIS[*cis*-(DIACETYLTETRACARBONYLMANGANESE)]ALUMINUM

$$3[CH_3C(O)]Mn(CO)_5 + 3CH_3Li + AlCl_3 \longrightarrow$$

$$[(OC)_4Mn(C(CH_3)O)_2]_3Al + 3LiCl$$

Procedure

A 100-mL, two-necked flask is removed from a drying oven (130°) and immedi-

*The checkers recorded these values with a Perkin-Elmer Model 337 spectrometer equipped with an expanded-scale recorder, calibrated with gaseous CO and DCl.

ately flushed well with prepurified nitrogen, after a gas inlet is attached to the side neck. The following procedures should be performed under a nitrogen atmosphere unless stated otherwise. The flask is charged with 1.0 g (4.2 mmole) of acetylpentacarbonylmanganese (see Sec. 11-A), 5 mL of freshly distilled diethyl ether, and a stirring bar. (■ **Caution.** *Metal carbonyl compounds are extremely toxic chemicals and should be handled in a good hood.*) The reaction flask is fitted with a septum cap on the center neck and is cooled to 0°. After 5 minutes 2.3 mL of a 2.06 *M* methyllithium solution (4.7 mmole) in diethyl ether is added with a syringe over a 10-minute period. During this time the reaction solution becomes a deep yellow and any solid previously present dissolves. The reaction solution is allowed to stir at 0° for an additional 45 minutes.

A stock solution of anhydrous aluminum chloride is prepared by weighing 0.38 g of anhydrous $AlCl_3$ in an 8-dram vial under nitrogen and dissolving the $AlCl_3$ in 10.0 mL of diethyl ether. To the cold reaction solution is added 5.0 mL of the $AlCl_3$ stock solution (1.4 mmole) by means of syringe over a 5-minute period. During this addition an off-white precipitate begins to form. The reaction solution is stirred at 0° for an additional 90 minutes. The precipitate is collected by filtration through a Schlenk frit[5] and washed twice with 1-mL portions of diethyl ether and dried at reduced pressure.*

The solid is transferred to another 100-mL, two-necked flask and is stirred with 50 mL of dichloromethane for 45 minues at 25°. During this time the solution becomes orange. This extractant mixture is filtered through a Schlenk frit,[5] and the solvent is removed under reduced pressure, affording 0.36 g (33%) of tris[*cis*-(diacetyltetracarbonylmanganese)] aluminum as an off-white solid. This solid is of sufficient purity for most purposes. Recrystallization from toluene at −20° affords single crystals. *Anal.* Calcd. for $C_{24}H_{18}AlMn_3O_{18}$: C, 36.66; H, 2.31; Al, 3.43, Mn, 20.96. Found: C, 36.50; H, 2.41; Al, 3.45; Mn, 20.66.

Properties

Tris[*cis*-(diacetyltetracarbonylmanganese)] aluminum is a pale-yellow solid that crystallizes as monoclinic needles. This solid decomposes rapidly at 265°, is air stable for at least 2 days at 25°, and has good to excellent solubility in most organic solvents. The infrared spectrum in cyclohexane† solution shows $\nu(C{\equiv}O)$ bands at 2065 (m), 1986 (s, sh), 1980 (s), 1963 (s) and a $\nu(C{\cdots}O)$ band at 1525 cm^{-1}. The ^1H NMR spectrum ($CDCl_3$ versus TMS) shows a singlet for the six equivalent methyl groups at $\tau 7.28$.

*The checkers conducted this filtration and subsequent steps in air with no reduction in yield.

†The checkers recorded these values with a Perkin-Elmer Model 337 spectrometer equipped with an expanded-scale recorder, calibrated with gaseous CO and DCl.

References

1. C. M. Lukehart, G. P. Torrence, and J. V. Zeile, *J. Am. Chem. Soc.*, **97**, 6903 (1975).
2. E. A. Shugam and L. M. Shkolnikova, *Dokl. Akad. Nauk S.S.S.R.*, **133**, 386 (1960).
3. C. M. Lukehart, G. P. Torrence, and J. V. Zeile, *Inorg. Chem.*, **15**, 2393 (1976).
4. Appendix, *Inorg. Synth.*, **12**, 317 (1970).
5. R. B. King, in *Organometallic Syntheses*, Vol. 1, J. J. Eisch and R. B. King (eds.), Academic Press Inc., New York, 1965 p. 56.
6. W. Beck, W. Hieber, and H. Tengler, *Chem. Ber.*, **94**, 862 (1961).
7. Ethyl Corporation British Patent 863, 277, (1957) *Chem. Abstr.*, **56**, 9969 (1962).

12. DODECACARBONYLTETRA-μ-HYDRIDO-*tetrahedro*-TETRARHENIUM

$$2Re_2(CO)_{10} + 2H_2 \text{ (1 atm)} \xrightarrow[150\text{-}160^\circ]{} Re_4H_4(CO)_{12} + 8CO$$

Submitted by J. R. JOHNSON and H. D. KAESZ*
Checked by B. F. G. JOHNSON†

The title compound was initially synthesized by the pyrolysis of $Re_3H_3(CO)_{12}$ at 190° in hydrocarbon solution.[1] Treatment of the complex with carbon monoxide at atmospheric pressure gradually converts it into higher carbonyls, as indicated in the reaction sequence below. At slightly elevated temperatures, the reaction is much faster and hydrogen is evolved. The suggestion that the reverse transformation might be possible led to the current synthesis.[2] The direct hydro-

$$Re_4H_4(CO)_{12} + 5CO \text{ (1 atm)} \xrightarrow{\text{cyclohexane}} Re_3H_3(CO)_{12} + ReH(CO)_5$$

$$ReH(CO)_5 \xrightarrow{60^\circ} \tfrac{1}{2}H_2 + \tfrac{1}{2}Re_2(CO)_{10}$$

genation of metal carbonyls is fairly general and may be used for the synthesis of $Re_4H_4(CO)_{12}$, $Ru_4H_4(CO)_{12}$, $Os_3H_2(CO)_{10}$, $Os_4H_4(CO)_{12}$, and $Re_3H_3(CO)_{12}$[2-4] Also it is adapted readily for the preparation of the corresponding deuterides.[3]

*Department of Chemistry, University of California, Los Angeles, CA 90024.
†University Chemical Laboratory, Cambridge CB2 1EW England.

Procedure

■ **Caution.** *Poisonous carbon monoxide evolved from this reaction and excess explosive hydrogen must be vented properly. All metal carbonyls should be considered to be toxic compounds.*

Rhenium carbonyl* (2.0 g, 3.07 mmole) is placed in a 250-mL, three-necked, round-bottomed flask containing dry deaerated decahydronaphthalene (150 mL) and a magnetic stirring bar. A fritted gas bubbler is connected to one neck of the flask, a reflux condenser to the second neck, and a thermometer (0-250°) to the third. A mercury bubbler is connected to the top of the condenser for a gas exit port. The solution is heated at 150-160° with magnetic stirring while hydrogen gas (99.9%) is slowly bubbled through. The initially colorless solution turns light yellow after 3 hours, but only $Re_2(CO)_{10}$ can be seen in the infrared spectrum of the carbonyl region. After about 7 hours the solution is light red and the carbonyl infrared region shows mainly $Re_2(CO)_{10}$ with some Re_3H_3-$(CO)_{12}$. After about 28 hours the solution is deep red and the carbonyl infrared region shows only $Re_4H_4(CO)_{12}$. A very thin rhenium mirror is sometimes present on the surfaces of the flask. The reaction mixture is then cooled to room temperature under nitrogen, since $Re_4H_4(CO)_{12}$ is somewhat air sensitive. The solution is chilled to 0° to crystallize any unreacted $Re_2(CO)_{10}$ and Re_3H_3-$(CO)_{12}$. The $Re_4H_4(CO)_{12}$ solution is filtered and dry, distilled, deaerated benzene (10 mL) is added to the filtrate. The solution is allowed to sit at room temperature for several hours under nitrogen during which time the decahydronaphthalene solvate of $Re_4H_4(CO)_{12}$ precipitates. The compound is collected by filtration and dried *in vacuo*. Typical yields based on Re are 0.80-0.95 g (43-51%). The dried compound may be recrystallized from hexane-dichloromethane at 0° yielding beautiful shiny crystalline platelets that have a greenish sheen. *Anal.* Calcd. for $Re_4H_4(CO)_{12}\cdot C_{10}H_{18}$: C, 21.59; H, 1.81. Found: C, 22.99; H, 1.60. When the red powder is dissolved in cyclohexane and reprecipitated with benzene, the product is a cyclohexane solvate. *Anal.* Calcd. for $Re_4H_4(CO)_{12}\cdot C_6H_{12}$: C, 18.49; H, 1.38. Found: C, 18.84; H, 1.39.

Properties

The compound $Re_4H_4(CO)_{12}$ is a very dark red (green by reflected light) crystalline compound that is somewhat air stable in the solid state but should be stored under nitrogen, especially when in solution. The infrared spectrum in the carbonyl region shows only two bands, 2042 and 1990 cm^{-1}. The ^1H NMR (CCl_4 solution) show a resonance at τ15.08 together with the resonances for the solvent of crystallization. The compound reacts slowly with carbon monoxide and instant-

*$Re_2(CO)_{10}$ may be purchased from either Strem Chemicals, Inc., Beverly MA or Pressure Chemicals, Pittsburgh, PA.

ly with any solvents containing donor atoms (acetone, diethyl ether, acetonitrile, etc.). Treatment of $Re_4H_4(CO)_{12}$ in cyclohexane solution in a heterogeneous reaction with solid sodium tetrahydroborate results in a very slow reaction (overnight) in which the red color and all traces of the carbonyl complex are observed to disappear from the hydrocarbon solvent. Removal of the cyclohexane and addition of acetone reveals by infrared spectrometry two carbonyl bands of the $[Re_4H_6(CO)_{12}]^{2-}$ ion (2000 and 1910 cm^{-1}) previously isolated from the reaction of $Re_2(CO)_{10}$ with $NaBH_4$.[5]

References

1. R. B. Saillant, G. Barcelo, and H. D. Kaesz, *J. Am. Chem. Soc.*, **92**, 5739 (1970).
2. H. D. Kaesz, S. A. R. Knox, J. W. Koepke, and R. B. Saillant, *Chem. Commun.*, **1971** 477.
3. S. A. R. Knox, J. W. Koepke, M. A. Andrews, and H. D. Kaesz, *J. Am. Chem. Soc.*, **97**, 3942 (1975).
4. $Re_3H_3(CO)_{12}$ is better prepared from $Re_2(CO)_{10}$ and $NaBH_4$; M. A. Andrews, S. W. Kirtley, and H. D. Kaesz, *Inorg. Synth.*, **17**, 66 (1977).
5. H. D. Kaesz, B. Fontal, R. Bau, S. Kirtley, and M. R. Churchill, *J. Am. Chem. Soc.*, **91**, 1021 (1969).

13. CARBONYL TRIMETHYLPHOSPHINE IRIDIUM(I) COMPLEXES

Submitted by J. A. LABINGER* and J. A. OSBORN†
Checked by A. DAVISON‡

Although a general route for the preparation of carbonylhalobis(*tert*-phosphine)-iridium(I) complexes has been reported,[1,2] it is somewhat inconvenient for complexes of highly basic ligands (e.g., trimethyl- or dimethylphenylphosphine), since the two-step sequence required[2] and the high solubility of the products (particularly for the $P(CH_3)_3$ complex) result in low yields.§ A more effici-

*Department of Chemistry, University of Notre Dame, Notre Dame, IN 46556.

†Institut Le Bel, Universite Louis Pasteur, 67000 Strasbourg, France.

‡Department of Chemistry, Massachusetts Institute of Technology, Cambridge, MA 02139.

§It has also been found that this general method, which involves reductive carbonylation of an iridium salt under CO in an alcoholic solvent followed by addition of phosphine ligand, is not always reliable. On occasion the carbonylation step is unsuccessful, leading to formation of a dark suspension instead of the expected clear yellow solution; addition of phosphine to the suspension affords no product. Although the reason for this erratic behavior is not clear, its frequency of occurrence appears to vary from one sample of starting iridium salt to another, suggesting that small amounts of impurities in some samples may be responsible.

ent synthesis for such complexes is desirable, since in a number of studies on oxidative addition reactions of iridium(I) complexes, the more easily obtained complexes of less basic ligands have been found to be insufficiently reactive.[3,4] The procedure described here employs a solid-state reaction to obtain the trimethylphosphine complex, thus affording yields substantially higher than those achieved in solution. The starting material is the analogous triphenylphosphine complex $IrCl(CO)[P(C_6H_5)_3]_2$; an excellent synthesis of this compound from readily available $IrCl_3 \cdot xH_2O$ has been reported.[5]

A. CARBONYLTETRAKIS(TRIMETHYLPHOSPHINE)IRIDIUM(I) CHLORIDE

$$IrCl(CO)(P(C_6H_5)_3)_2 + 4P(CH_3)_3 \longrightarrow [Ir(CO)(P(CH_3)_3)_4]Cl + 2P(C_6H_5)_3$$

Procedure

■ **Caution.** *Trimethylphosphine is toxic and spontaneously flammable in air. Manipulations involving the free ligand should be performed using vacuum-line techniques.*[6]

A flask containing 5.6 g (7.1 mmole) of carbonylchlorobis(triphenylphosphine)iridium(I)[5] suspended in 50 mL of benzene and equipped with a magnetic stirrer is attached to a vacuum line, frozen in liquid nitrogen, and evacuated. Trimethylphosphine* (2.2 g, 28.4 mmole) is condensed into the flask, which is then closed off from the line and allowed to warm to room temperature with stirring. The resulting suspension is filtered (in air) and the solid is washed with benzene to remove traces of yellow color and then dried in vacuum. The yield of $[Ir(CO)(P(CH_3)_3)_4]Cl$ is 4.0 g (98%).

Properties

The complex is a white solid, insoluble in nonpolar solvents but soluble in polar solvents such as dichloromethane, ethanol, and even water. The solid is stable in air for fairly long periods. Very slow decomposition, probably by reaction with atmospheric water, is observed over several days. Solutions of the complex are pale yellow, smell of free trimethylphosphine (*toxic*), and exhibit a very broad ^1H NMR signal, indicating that reversible dissociation of phosphine and coordination of chloride takes place in solution. Treatment with $Na[B(C_6H_5)_4]$ in ethanol affords the tetraphenylborate salt, which shows a sharp, complex multi-

*The checker recommends that the trimethylphosphine be generated by thermal decomposition of the silver iodide complex, $[AgI(P(CH_3)_3)]_4$, which can be obtained readily from the free phosphine ligand[8] and is easier to handle.

plet in the NMR and may be recrystallized from dichloromethane-ethanol without decomposition. *Anal.* Calcd. for $C_{37}H_{56}BIrOP_4$: C, 52.7; H, 6.70. Found: C, 52.79; H, 6.51. The chloride and the tetraphenylborate salts show a strong infrared peak at about 1900 cm^{-1}. In contrast to the closely related compound $[Ir(CO)(diphos)_2]Cl$, which loses CO on heating in vacuum,[7] $[Ir(CO)((CH_3)_3)_4]Cl$ loses $P(CH_3)_3$ (see below). [diphos = ethylenebis(diphenylphosphine)]

B. CARBONYLCHLOROBIS(TRIMETHYLPHOSPHINE)IRIDIUM(I)

$$[Ir(CO)(P(CH_3)_3)_4]Cl \xrightarrow[-2P(CH_3)_3]{\Delta} IrCl(CO)(P(CH_3)_3)_2$$

Procedure

A sublimation apparatus (with a water-cooled probe) is charged with 4.0 g of carbonyltetrakis(trimethylphosphine)iridium(I) chloride, evacuated, and heated to 130°. At first a white film appears on the cold finger,* followed by growth of bright yellow crystals. After nearly all the solid has sublimed, the product is scraped from the cold finger (this may be done in air, but exposure should be kept to a minimum) and resublimed once or twice until the sublimate consists of pure yellow crystals with no visible white contaminant. The yield is 2.4 g (87%).† *Anal.* Calcd. for $C_7H_{18}ClIrOP_2$: C, 20.6; H, 4.46. Found: C, 21.10; H, 4.55.

Properties

The infrared (CO stretch at 1940 cm^{-1}) and ^1H NMR (triplet at δ 1.3) spectral parameters for $IrCl(CO)(P(CH_3)_3)_2$ prepared by this route are identical to those previously reported.[2] The solid may be handled in air for brief periods without visible effect; however, it should be noted that very small amounts of oxygen

*This white contaminant is $[Ir(CO)(P(CH_3)_3)_4]Cl$, presumably formed by reaction of $IrCl(CO)(P(CH_3)_3)_2$ with a small amount of liberated $P(CH_3)_3$ which cocondenses on the cold finger.

†Since $IrCl(CO)(P(C_6H_5)_3)_2$ can be prepared in better than 90% yield from $IrCl_3$ or other iridium salts,[5] the overall sequence from $IrCl_3$ to $IrCl(CO)(P(CH_3)_3)_2$ here gives a yield of 75-80%, compared to the 17% overall yield achieved previously.[2] While this procedure does require the use of excess $P(CH_3)_3$, the excess is liberated in the second step and can be collected for reuse. The checker recommends incorporation of a liquid-nitrogen-cooled trap into the vacuum line used during sublimation, which not only allows recovery of the liberated $P(CH_3)_3$, but also reduces the amount of white contaminant in the sublimate, allowing pure product to be obtained after only one sublimation.

may have pronounced effects upon reactions of the complex.[3] Hence it is probably preferable to handle the purified product under an inert atmosphere. Prolonged exposure of the solid to air (or brief exposure of solutions) results in the growth of a new infrared peak at 2000 cm^{-1}, indicating formation of the oxygen adduct, accompanied by darkening and decomposition. Partially oxidized samples may be purified by resublimation.

References

1. J. Chatt, N. P. Johnson, and B. L. Shaw, *J. Chem. Soc. (A)*, **1967**, 604.
2. A. J. Deeming and B. L. Shaw, *J. Chem. Soc. (A)*, **1968**, 1887.
3. J. S. Bradley, D. E. Connor, D. Dolphin, J. A. Labinger, and J. A. Osborn, *J. Am. Chem. Soc.*, **94**, 4043 (1972).
4. M. R. Churchill, J. J. Hackbarth, A. Davison, D. D. Traficante, and S. S. Wreford, *J. Am. Chem. Soc.*, **96**, 4041 (1974).
5. J. P. Collman, C. T. Sears, and M. Kubota, *Inorg. Synth.*, **11**, 101 (1968); J. P. Collman and J. W. Kang, *J. Am. Chem. Soc.*, **89**, 844 (1967).
6. D. F. Shriver, "The Manipulation of Air-sensitive Compounds," McGraw-Hill Book Company, New York, 1969.
7. L. Vaska and D. L. Catone, *J. Am. Chem. Soc.*, **88**, 5324 (1966).
8. R. T. Markham, E. A. Dietz, and D. R. Martin, *Inorg. Synth.*, **16**, 153 (1976).

Chapter Three

OTHER COORDINATION COMPOUNDS

14. COMPLEXES OF COBALT CONTAINING AMMONIA OR ETHYLENEDIAMINE

HEXAAMMINECOBALT(III) SALTS

Submitted by ROBERT D. LINDHOLM*
Checked by DANIEL E. BAUSE†

Hexaamminecobalt(III) salts are generally prepared[1-4] by the oxidation of ammoniacal cobalt(II) solutions by either H_2O_2 or O_2 in the presence of a catalyst. The preparation procedure[4] most often employed involves the aerial oxidation of an ammoniacal cobalt(II) solution in the presence of a carbon catalyst. The new procedure employs the same reaction conditions but utilizes a nonaqueous solvent to simplify the synthesis and to prepare the acetate salt, which is very soluble in water.

*Research Laboratories, Eastman Kodak Company, Rochester, NY 14650.
†Department of Chemistry, University of Pittsburgh, Pittsburgh, PA 15260.

A. HEXAAMMINECOBALT(III) ACETATE

$$4Co(C_2H_3O_2)_2 \cdot 4H_2O + 4NH_4C_2H_3O_2 + 20\ NH_3 + O_2 \longrightarrow 4[Co(NH_3)_6](C_2H_3O_2)_3 + 18H_2O$$

Procedure

Twenty-five grams (0.1 mole) of cobalt(II) acetate tetrahydrate and 7.7 g (0.1 mole) of ammonium acetate are dissolved in 500 mL of technical grade methanol. Four grams of granular activated carbon is added to the solution, and NH_3 and air are bubbled through the mixture for 2 hours. The resulting mixture is then filtered through a Büchner filter to remove the carbon catalyst and the filtrate is diluted at least 4:1 with acetone to precipitate the product. The resulting $[Co(NH_3)_6](C_2H_3O_2)_3$ is filtered, washed with diethyl ether, and air-dried. The yield is 30 g (90%). The checker obtained 19.7 g (58.3%). *Anal.* Calcd. for $[Co(NH_3)_6](C_2H_3O_2)_3$: Co, 17.39; C, 21.30; H, 8.05; N, 24.85. Found: Co, 17.2; C, 20.7; H, 7.9; N, 24.1.

B. HEXAAMMINECOBALT(III) CHLORIDE

$$4Co(C_2H_3O_2)_2 \cdot 4H_2O + 4NH_4C_2H_3O_2 + 20\ NH_3 + O_2 \longrightarrow 4[Co(NH_3)_6](C_2H_3O_2)_3 + 18H_2O \quad (1)$$

$$[Co(NH_3)_6](C_2H_3O_2)_3 + 3HCl \longrightarrow [Co(NH_3)_6]Cl_3 + 3HC_2H_3O_2 \quad (2)$$

Procedure

Twenty-five grams (0.1 mole) of cobalt(II) acetate tetrahydrate and 7.7 g (0.1 mole) of ammonium acetate are dissolved in 500 mL of technical grade methanol. Four grams of granular activated carbon is added to the solution, and NH_3 and air are bubbled through the mixture for 2 hours. The resulting mixture is then filtered through a Büchner filter to remove the carbon catalyst and the filtrate is treated with 25 mL of concentrated HCl. The resulting $[Co(NH_3)_6]Cl_3$ is filtered, washed with ethanol and then diethyl ether, and air-dried. The yield is 20 g (80%). The checker obtained 14.4-16.1 g of a red-brown product. This crude product was recrystallized by dissolving the sample in a minimum of hot water (80°) followed by addition of 20 mL of concentrated HCl. The desired yellow-orange product was filtered, washed, and dried as before. The yield was 8.5-9.7 g. *Anal.* Calcd. for $[Co(NH_3)_6]Cl_3$: Co, 22.05; H, 6.80; N, 31.45. Found: Co, 22.15; H, 6.7; N, 31.1.

Properties

Hexaamminecobalt(III) ion exhibits two absorption maxima at 475 and 340 nm with extinction coefficients of 58 and 49, respectively. The solubility of the chloride form in water at 20° is 0.26 moles/L. The solubility of the acetate salt is about 1.9 moles/L at 20°. In addition, it has fair solubility in aliphatic and aromatic alcohols.

References

1. M. Smith and M. Smith, *J. Am. Chem. Soc.*, **121**, 1970 (1922).
2. W. Biltz, *Z. Anorg. Chem.*, **83**, 178 (1914).
3. S. Jørgensen, *Z. Anorg. Chem.*, **17**, 457 (1898).
4. J. Bjerrum and J. McReynolds, *Inorg. Synth.*, **2**, 216 (1946).

cis- and *trans*-[TETRAAMMINEDINITROCOBALT(III)] NITRATE

Submitted by GEORGE B. KAUFFMAN,* STEVEN F. ABBOTT,* STEPHEN E. CLARK,* JOHN M. GIBSON,* and ROBIN D. MEYERS*
Checked by GLENN J. NICHOLS,† JEFFREY H. WENGROVIUS,† and JOHN C. BAILAR, JR.‡

The tetraamminedinitrocobalt(III) salts represent the second longest known case of geometric isomerism among coordination compounds. The reddish-yellow trans compounds were first prepared in 1875 by air oxidation of a solution of cobalt(II) chloride containing ammonia, sodium nitrite, and ammonium chloride by Wolcott Gibbs,[1] who called them *croceo* salts. The brownish-yellow cis compounds were first prepared in 1894 by reaction of tetraamminecarbonatocobalt(III) salts with sodium nitrite by Jørgensen,[2] who called them *flavo* salts. The two series correspond completely to the bis(ethylenediamine)dinitrocobalt(III) salts.

*California State University, Fresno, Fresno, CA 93740.
†The Colorado College, Colorado Springs, CO 80903.
‡Department of Chemistry, University of Illinois, Urbana, IL 61803.

In both series, which are not interconvertible, the two nitro groups are "masked" within the coordination sphere and are not removed during metathetical reactions. Further evidence for the strong bonding of the nitro groups to the cobalt atom is the existence of the complex in solution as a monopositive ion, as demonstrated by measurements of freezing point,[3,4] conductance,[4-10] flocculating power,[11] and transport number.[12,13] Because of the difference in the reactivity of the nitro groups toward hydrochloric acid in the two series, Jørgensen regarded them as structural isomers and considered flavo salts as nitrito (Co—ONO) compounds and croceo salts as nitro ($Co—NO_2$) compounds. Werner,[14-16] on the basis of conversions to compounds of known configuration, regarded them as stereoisomers and assigned them the configurations that are accepted today. The syntheses given below are based on the work of Jørgensen (cis[17,18] and trans[19,20]).

C. *cis*-[TETRAAMMINEDINITROCOBALT(III)] NITRATE

$$[Co(CO_3)(NH_3)_4]NO_3 \cdot 0.5H_2O + 2HNO_3 + 2NaNO_2 \longrightarrow$$
$$cis\text{-}[Co(NH_3)_4(NO_2)_2]NO_3 + 2NaNO_3 + CO_2 + 1.5\ H_2O$$

Procedure

Ten grams of tetraammine(carbonato)cobalt(III) nitrate 0.5 hydrate[21] (0.0388 mole) is added in small portions with stirring (■ **Caution.** *Effervescence*) to a mixture of 100 mL of water and 6.0 mL of concentrated nitric acid contained in a 250-mL beaker. To the resulting deep-red solution of *cis*-[tetraammine-diaquacobalt(III)] nitrate is added, with stirring, 20.0 g of sodium nitrite in small portions, whereupon fumes of nitrogen oxides are evolved (*Hood!*). The effervescence is allowed to subside before each new portion of sodium nitrite is added. The solution is heated on a steam bath until the color changes to brownish yellow (about 10 min). It is then cooled *immediately to room temperature* in an ice bath, and 80 mL of concentrated nitric acid is added with stirring, whereupon nitrogen oxides are again evolved (*Hood!*).* When precipitation of the brownish-yellow crystals appears complete (about ½ hour), the crude product is collected by suction filtration on a 7-cm Büchner funnel, washed with three 20-mL portions of ice-cold 1:1 nitric acid, and three 20-mL portions of ice-cold 95% ethanol, and then air-dried. The yield of crude product is 9.50 g 87.2%).

*The acid should be added initially in small portions to prevent loss of material by excessive foaming.

The product is purified by dissolving it in a minimum volume (about 45 mL) of boiling 1% acetic acid, cooling in an ice bath, and collecting the brownish-yellow crystals on a 7-cm Büchner funnel. The product is washed with three 20-mL portions of an ice-cold 1:1 ethanol-water mixture and three 20-mL portins of ice-cold 95% ethanol, air-dried, and then dried in an oven at 50° for 1 hour. The yield is 8.30 g (76.2%). *Anal.* Calcd. for $[Co(NH_3)_4(NO_2)_2]NO_3$: Co, 20.97; N, 34.88. Found: Co, 20.82; N, 35.06.

D. *trans*-[TETRAAMMINEDINITROCOBALT(III)] NITRATE

$$4CoCl_2 \cdot 6H_2O + 4NH_4Cl + 8NaNO_2 + 12NH_3 + O_2 \longrightarrow 4\ trans\text{-}[Co(NH_3)_4(NO_2)_2]Cl + 8NaCl + 26H_2O$$

$$trans\text{-}[Co(NH_3)_4(NO_2)_2]Cl + NH_4NO_3 \longrightarrow trans\text{-}[Co(NH_3)_4(NO_2)_2]NO_3 + NH_4Cl$$

Procedure

Twenty-seven grams (0.391 mole) of sodium nitrite and 20.0 g (0.374 mole) of ammonium chloride are dissolved with stirring in 150 mL of water contained in a 500-mL Erlenmeyer filter flask. To this solution are added with stirring 20 mL of concentrated (28%) aqueous ammonia and a solution of 18.0 g (0.0757 mole) of cobalt(II) chloride hexahydrate dissolved in 50 mL of water. The flask is closed with a stopper carrying an inlet tube (at least 10 mm od) reaching to the bottom of the flask, and by means of an aspirator a rapid current of air is drawn through the solution for a period of 5 hours. At first moisture condenses on the outside walls of the flask, which becomes cold. During the oxidation, the color of the solution changes from yellowish-red to dark yellowish-brown, and a light orange-brown precipitate forms. The flask is then stoppered and allowed to stand for 24 hours in the dark. (The product is somewhat photosensitive.)

The orange-brown precipitate is collected by suction filtration on a 7-cm Büchner funnel, washed with 25 mL of ice water, and tested for the absence of any contaminating pentaamminenitrocobalt(III) chloride, which may be present in small amounts. Twenty-five milliliters of water is poured onto the precipitate and allowed to drain without suction into the filter flask, which has been emptied and rinsed with water. Suction is then applied, and 10 mL of 0.5 *M* ammonium oxalate is added to the yellowish-brown filtrate. No precipitate or only turbidity within 30 minutes indicates that the amount of pentaamminenitro complex present is negligible. If the test is positive, the precipitate should be washed as above with additional 10-mL portions of water until the filtrate gives no reaction with the oxalate. The precipitate is then washed with three 20-mL portions of 95% ethanol and air-dried.

The precipitate (about 12.5-13.0 g), consisting of approximately equal amounts of *trans*-[tetraamminedinitrocobalt(III)] chloride and nitrate (formed by oxidation of excess nitrite), is converted completely to the sparingly soluble nitrate and recrystallized by placing it in a 500-mL Erlenmeyer flask, adding 400 mL of boiling water containing 0.5 mL of glacial acetic acid, shaking the mixture well, and heating *gently* until no more solid dissolves (■ **Caution.** *Much of the nitrate remains undissolved. Prolonged heating should be avoided.*) Fifty-five grams of ammonium nitrate is dissolved with stirring in the hot mixture, which is immediately cooled under a stream of tap water and then in an ice bath in the dark (about 4 hours). The deposited crystals are collected by suction filtration on a 7-cm Büchner funnel, washed with three 20-mL portions of 95% ethanol and then with three 20-mL portions of diethyl ether, and air-dried. The product is then dried at 100° for 1½ hours. The yield of golden-orange crystals is 12.5 g (58.7%). The product should be stored in a tightly stoppered dark bottle and kept in the dark. *Anal.* Calcd. for $[Co(NH_3)_4(NO_2)_2]$-NO_3: Co, 20.97; N, 34.88. Found: Co, 20.43; N, 34.71.

Properties

The *cis*-[tetraamminedinitrocobalt(III)] complexes are more soluble than the corresponding trans complexes.[2] The two series can be distinguished by the action of various reagents. The cis complexes give blue *cis*-$[CoCl_2(NH_3)_4]Cl$ at low temperature or the green *trans*-$[CoCl_2(NH_3)_4]Cl$ at room temperature with concentrated hydrochloric acid, a precipitate on treatment with ammonium oxalate or hexafluorosilicic acid,[2,19] and the red *cis*-$[Co(H_2O)_2(NH_3)_4](NO_3)_3$ on treatment with 50% nitric acid.[22] In the catalytic decomposition of hydrogen peroxide, they are less active than the corresponding trans complexes. In aqueous solution aquation occurs quickly.

Infrared absorption bands are observed as follows: (KBr, 0.5%): 3280 (vs), 3170 (vs), 1615-1620 (m), 1555 (m), 1420 (s), 1380 (vs), 1340 (m), 1310 (s), 1295 (s), 840 (vw), 828 (w), 602 (w), 580 (w), 505 (vw), and 460 (w) cm^{-1}, in close agreement with published data for the chloride.[23,24]

Ultraviolet-visible absorption bands (in deionized water) appear at 440 (w), 322 (s), and 234 (s) nm, in close agreement with published data for the chloride.[25]

The *trans*-[tetraamminedinitrocobalt(III)] complexes give the brownish-red *trans*-$[CoCl(NH_3)_4(NO_2)]Cl$ with concentrated hydrochloric acid, no precipitate with ammonium oxalate or hexafluorosilicic acid,[2,19] and the brown-yellow *trans*-$[Co(H_2O)(NH_3)_4NO_2](NO_3)_2$ with 50% nitric acid.[22] At 18° the solubility of *trans*-$[Co(NH_3)_4(NO_2)_2]NO_3$ is 0.0119 mole/L.

Infrared bands (KBr, 0.5%) appear at 3290 (s), 3180 (s), 1620 (w), 1412 (m), 1400 (m), 1380 (s), 1320 (s), 1270 (s), 835 (w), 815 (m), 620 (vw), and 500 (vw) cm^{-1}, in close agreement with published data for the chloride.[23,24]

Ultraviolet-visible absorption bands (in deionized water) appear at 440 (w), 346 (s), and 251 (s) nm, in close agreement with published data for the chloride.[25]

E. *trans*-[DICHLOROBIS(ETHYLENEDIAMINE)COBALT(III)] NITRATE

$$trans\text{-}[Co(en)_2(NO_2)_2]NO_3 + 2HCl \longrightarrow trans\text{-}[CoCl_2(en)_2]NO_3 + NO_2 + NO + H_2O$$

Submitted by JACK ZEKTZER*
Checked by ROBERT E. HERMER†

A previously published procedure[26] for the preparation of *trans*-$[CoCl_2(en)_2]Cl$ gave an overall yield of 35% based on cobalt(II) chloride. The procedure must be followed exactly to avoid contamination of the product with $[Co(en)_3]Cl_3$. The procedures of Springborg and Schäffer[27] give yields of 75% but require 3-4 days for the crystallization of the product. The procedure set forth below gives overall yields of 70% and only requires 3-4 hours beginning with $[Co(en)_2(NO_2)_2]$-NO_3, or a total of 12 hours beginning with $Co(NO_3)_2 \cdot 6H_2O$.

Procedure

A mixture of 275.3 g (0.826 mole) of *trans*-$[Co(en)_2(NO_2)_2]NO_3$ prepared by the procedure of Holtzclaw et al.,[28] 800 mL of 12 M HCl, and 400 mL of water is heated on a hot plate until the evolution of NO_2 ceases (about 2-2.5 hr). *The use of a hood is mandatory.* During the heating, the color of the solution changes from orange to black-green. After the evolution of NO_2 ceases, *trans*-$[CoCl_2$-$(en)_2]NO_3$ precipitates as small green plates. The suspension is cooled in an ice bath and filtered. The precipitate is washed with methanol and air-dried. Yield

*6308½ Eighteenth St., N. E., Seattle, WA 98115.
†Department of Chemistry, University of Pittsburgh, Pittsburgh, PA 15260.

207-215 g (80-84%) based on $[Co(en)_2(NO_2)_2]NO_3$. *Anal.* Calcd. for $[CoCl_2(en)_2]NO_3$: C, 15.39; H, 5.13; N, 22.43. Round: C, 15.30; H, 4.94; N, 22.15.

Properties

The absorption spectral data agree well with those reported for the chloride salt.[27] The nitrate salt is sparingly soluble in water (0.346 g in 100 mL of water at 7°)[29] and can be recrystallized from hot water. The chloride salt[29] is much more soluble in water.

References

1. W. Gibbs, *Proc. Am. Acad. Sci.,* **10**, 1, 7 (1875).
2. S. J. Jørgensen, *Z. Anorg. Chem.,* **5**, 159, 189 (1894).
3. A. Werner and C. Herty, *Z. Phys. Chem.,* **38**, 331 (1901).
4. W. D. Harkins, R. E. Hall, and W. A. Roberts, *J. Am. Chem. Soc.,* **38**, 2643 (1916).
5. A. Werner and A. Miolati, *Z. Phys. Chem.,* **12**, 48 (1893).
6. J. Meyer and H. Moldenhauer, *Z. Anorg. Allgem. Chem.,* **118**, 28 (1921).
7. Y. Shibata, *J. Coll. Sci. Imp. Univ. Tokyo,* **37**, Art. 8 (1916).
8. N. Dhar, *Z. Anorg. Chem.,* **80**, 57 (1913).
9. A. B. Lamb and V. Yngve, *J. Am. Chem. Soc.,* **43**, 2360 (1921).
10. H. J. S. King, *J. Chem. Soc.,* **127**, 2109 (1925).
11. K. Matsuno, *J. Coll. Sci. Imp. Univ. Tokyo,* **41**, Art. 11 (1921).
12. R. Lorenz and I. Posen, *Z. Anorg. Chem.,* **95**, 345 (1916); *ibid.,* **96**, 86 (1916).
13. E. N. Gapon, *Z. Anorg. Allgem. Chem.,* **168**, 127 (1928).
14. A. Werner, *Z. Anorg. Chem.,* **8**, 182 (1895).
15. A. Werner, *Chem. Ber.,* **34**, 1705 (1901).
16. A. Werner, *Chem. Ann.,* **386**, 24, 29, 37 (1912).
17. S. M. Jørgensen, *Z. Anorg. Chem.,* **5**, 162 (1894); **17**, 473 (1898).
18. H. Biltz and W. Biltz (adapted from the German by W. T. Hall and A. A. Blanchard), *Laboratory Methods of Inorganic Chemistry,* John Wiley Inc., New York, 1928, p. 179.
19. S. M. Jørgensen, *Z. Anorg. Chem.,* **17**, 468, 471 (1898).
20. W. G. Palmer, *Experimental Inorganic Chemistry,* Cambridge University Press, Cambridge, England, 1970, p. 537.
21. G. Schlessinger, *Inorg. Synth.,* **6**, 173 (1960).
22. R. G. Yalman and T. Kuwana, *J. Phys. Chem.,* **59**, 298 (1955).
23. I. R. Beattie and H. J. V. Tyrrell, *J. Chem. Soc.,* **1956**, 2849.
24. K. Nakamoto, J. Fujita, and H. Murata, *J. Am. Chem. Soc.,* **80**, 4817 (1958).
25. F. Basolo, *J. Am. Chem. Soc.,* **72**, 4393 (1950).
26. J. C. Bailar, Jr., *Inorg. Synth.,* **2**, 222 (1946).
27. J. Springborg, and C. E. Schäffer, *Inorg. Synth.,* **14**, 63 (1974).
28. H. F. Holtzclaw, Jr., D. P. Sheetz, and B. D. McCarthy, *Inorg. Synth.,* **4**, 177 (1953).
29. S. M. Jorgensen, *J. Prakt. Chem.,* **39**, (2) 23 (1889).

15. TETRAAMMINE AND BIS(ETHYLENEDIAMINE) COMPLEXES OF CHROMIUM(III) AND COBALT(III)

Submitted by JOHAN SPRINGBORG* and CLAUS ERIK SCHÄFFER†
Checked by MICHAEL L. WILSON,‡ J. MARCUS WHARTON,‡ and WILLIAM E. HATFIELD‡

Methods for the preparation and purification of salts of the dimeric complex ions di-μ-hydroxo-bis[tetraamminechromium(III)] and di-μ-hydroxo-bis[bis-(ethylenediamine)chromium(III)] and of the two corresponding cobalt(III) species are presented. The two ammine complex dimers are isolated as bromide and perchlorate salts. The two ethylenediamine complexes are isolated as dithionate, bromide, chloride, and perchlorate salts. All four dimers have been obtained by heating the corresponding *cis*-aquahydroxo complexes as the dithionate salts.

Methods for preparation of *cis*-[tetraammineaquachlorochromium(III)] sulfate, *cis*-[tetraammineaquahydroxochromium(III)] dithionate monohydrate, and *cis*-[tetraammineaquahydroxocobalt(III)] dithionate monohydrate as nearly pure intermediates are given. The preparation of crude *cis*-[aquabis(ethylenediamine)-hydroxochromium(III)] dithionate and a method for obtaining the pure salt are given. These *cis*-aquahydroxo salts have been used to prepare the following pure *cis*-diaqua salts; *cis*-[tetraamminediaquachromium(III)] perchlorate, *cis*-[tetra-amminediaquacobalt(III)] perchlorate, and *cis*-[diaquabis(ethylenediamine)-chromium(III)] bromide dihydrate.

The methods presented here are modifications of those given in the literature. In most of the preparations given below, the crude products are almost pure and are obtained in high yields. The crude salts have been purified by reprecipitation, and the visible absorption spectra, besides elemental analysis, have been used as a criterion for the purity. When it is stated that a sample is pure, this means the positions of the maxima and minima of the spectrum remained constant upon a further reprecipitation and that a deviation of less than 1% in the molar absorptivity (ϵ) (for both maxima and minima) was found between the two crops.

*Chemistry Department, Royal Veterinary and Agricultural Univeristy, Thorvaldsensvej 40, DK-1871 Copenhagen V., Denmark.
†Chemistry Department I (Inorganic Chemsitry), University of Copenhagen, H. C. Ørsted Institute, Universitetsparken 5, DK-2100 Copenhagen Ø., Denmark.
‡Department of Chemistry, University of North Carolina, Chapel Hill, NC 27514.

TABLE I. Optical absorption Data for Solution of Diaqua and Dihydroxo Complexes

Molar absorbances ϵ (for dinuclear complexes per two metal atoms) are given in L $mole^{-1}$ cm^{-1} and wavelength λ in nanometers. Minima are included in the table, since these are crucial in criteria of purity. The (ϵ, λ) values given for solutions of salts of the $[(en)_2Cr(OH)_2Cr(en)_2]^{4+}$ cation represent the equilibrium mixture between the di-μ-hydroxo and the mono-μ-hydroxo complex at 20° (see text).

Compound	Medium	$(\epsilon, \lambda)_{max}$		$(\epsilon, \lambda)_{min}$		$(\epsilon, \lambda)_{shoulder}$
cis-$[Cr(NH_3)_4(H_2O)_2](ClO_4)_3$	0.12 M $HClO_4$	(36.0, 496)[a]	(26.4, 367)[b]	(8.0, 419)		
	0.4 M NH_3	(40.5, 531)[a]	(41.3, 383)[b]	(18.3, 455)		
cis-$[Co(NH_3)_4(H_2O)_2](ClO_4)_3$	0.12 M $HClO_4$	(52.8, 504)[a]	(43.0, 355)[b]	(6.9, 413)	(3.1, 292)	
	0.4 M NH_3	(67.6, 528)[a]	(73.9, 377)[b]	(11.8, 441)	(17.5, 324)	
cis-$[Cr(en)_2(H_2O)_2]Br_3 \cdot 2H_2O$	0.12 M $HClO_4$	(65.6, 486)[a]	(42.0, 369)[b]	(16.6, 416)	(0.8, 290)	
	0.10 M NaOH	(62.1, 530)[a]	(65.7, 379)[b]	(20.6, 445)	(0.7, 295)	
cis-$[Cr(en)_2(H_2O)(OH)]S_2O_6$	0.12 M $HClO_4$	(65.8, 486)[a]	(41.9, 369)[b]	(16.5, 416)	(0.8, 290)	
	0.10 M NaOH	(62.4, 530)[a]	(65.8, 379)[b]	(20.7, 445)	(1.3, 295)	
$[(NH_3)_4Cr(OH)_2Cr(NH_3)_4]Br_4 \cdot 4H_2O$	0.012 M $HClO_4$	(126.7, 535)[a]	(68.9, 386)[b]	(19.4, 441)	(4.7, 311)	(15.9, 340) (7.60, 322)[d] (37.9, 278)
$[(NH_3)_4Cr(OH)_2Cr(NH_3)_4](ClO_4)_4 \cdot 2H_2O$	0.012 M $HClO_4$	(126.5, 535)[a]	(68.1, 386)[b]	(19.0, 441)	(3.5, 311)	(14.1, 340) (6.3, 322)[d] (33.6, 278)
$[(en)_2Cr(OH)_2Cr(en)_2]Br_4 \cdot 2H_2O$	0.012 M $HClO_4$	(178.8, 528)[a]	(115.2, 384)[b]	(34.3, 439)	(7.1, 311)	(38.6, 285)[d]
$[(en)_2Cr(OH)_2Cr(en)_2]Cl_4 \cdot 2H_2O$	0.012 M $HClO_4$	(177.4, 528)[a]	(115.2, 384)[b]	(34.5, 439)	(5.6, 313)	(36.2, 285)[d]
$[(en)_2Cr(OH)_2Cr(en)_2](ClO_4)_4$	0.012 M $HClO_4$	(177.8, 528)[a]	(115.1, 384)[b]	(33.8, 439)	(6.2, 313)	(37.9, 285)[d]

$[(NH_3)_4Co(OH)_2Co(NH_3)_4]Br_4 \cdot 4H_2O$	0.012 M $HClO_4$	(195.1, 528)[a]	(2620, 287)[c]	(47.5, 441)	(1960, 263)	(191, 370)[b]
$[(NH_3)_4Co(OH)_2Co(NH_3)_4](ClO_4)_4 \cdot 2H_2O$	0.012 M $HClO_4$	(194.8, 528)[a]	(2650, 287)[c]	(46.2, 441)	(1890, 263)	(187, 370)[b]
$[(en)_2Co(OH)_2Co(en)_2]Br_4 \cdot 2H_2O$	0.012 M $HClO_4$	(303, 526)[a]	(3030, 290)[c]	(71.8, 443)	(2600, 272)	(248, 380)[b]
$[(en)_2Co(OH)_2Co(en)_2]Cl_4 \cdot 5H_2O$	0.012 M $HClO_4$	(304, 526)[a]	(2940, 290)[c]	(72.6, 443)	(2570, 272)	(249, 380)[b]
$[(en)_2Co(OH)_2Co(en)_2](ClO_4)_4$	0.012 M $HClO_4$	(304, 526)[a]	(29.20, 290)[c]	(71.9, 443)	(2500, 272)	(247, 380)[b]

[a]First spin-allowed ligand field absorption band (cubic parentage),

$^4A_{2g} \rightarrow {}^4T_{2g}$ (O_h) for chromium(III)
$^1A_{1g} \rightarrow {}^1T_{1g}$ (O_h) for cobalt(III)

[b]Second spin-allowed ligand field absorption band (cubic parentage),

$^4A_{2g} \rightarrow {}^4T_{1g}$ (O_h) for chromium(III)
$^1A_{1g} \rightarrow {}^1T_{2g}$ (O_h) for cobalt(III)

[c]Unassigned band characteristic of the di-μ-hydroxocobalt(III) systems.
[d]Unassigned band caused by the weak exchange interaction between the two chromium(III) ions within the bridged complexes.

For some of the compounds the spectrum changed with time. A linear extrapolation of the spectrum back to the time of dissolution was then made. The corrections were never greater than 1%. In Table I the $(\epsilon, \lambda)_{extremum}$ values have been collected and are those obtained for the sample precipitated once more than that which is stated as pure.

■ **Caution.** *Mechanical handling and heating of perchlorates represents a potential danger because of the reducing character of coordinated ammonia or amines. It is often not realized that a glass rod scraped across a sintered-glass filter may create an exorbitant local pressure, which is likely to act as a shock. Therefore, the use of a rod of soft polyethylene is strongly recommended in the handling of perchlorates. Also, washing of solids with alcohol or other organic solvents should be done with caution. However, we have never experienced an explosion with the present compounds.*

The mononuclear cations of chromium(III) and cobalt(III) of the tetraammine and the bis(ethylenediamine) series belong to a class of complexes that has been most extensively studied. The present complexes will often prove useful as starting materials for preparing complexes within their class, and reactions with these complexes are likely to form the subject of many future investigations.

The two main reasons that the bridged complexes prepared here are of particular interest are as follows:

1. They represent bridged structures of known constitution whose properties may throw light upon the most common type of hydrolysis products of aqua ions within the whole periodic system. In this connection it is noteworthy that the aquation behavior of the di-μ-hydroxochromium(III) complex, *meso*-$[(en)_2Cr(OH)_2Cr(en)_2]^{4+}$ appears to be significantly different from that of the corresponding cobalt(III) complex.[1-3]

2. The chromium(III) complexes represent the most simple systems in which magnetic interactions between paramagnetic ions can be studied. The cobalt(III) systems may serve as matrixes in which the intermolecular interactions between the paramagnetic ions can be diminished. In these connections it is noteworthy that a mixed dihydroxo complex, $(-)_{589}$-Λ,Δ-$[(en)_2Cr(OH)_2Co(en)_2]^{4+}$, has been described recently.[4,5] This compound may serve as a model compound that represents the magnetic properties of the individual chromium(III) ion of the di-μ-hydroxochromium(III) complex in the absence of a pair interaction.

A. *cis*-[TETRAAMMINEAQUACHLOROCHROMIUM(III)] SULFATE

$$Cr_2(SO_4)_3 + 12NH_3 + 2NH_4Cl + 4H_2O \longrightarrow$$

$$2[Cr(NH_3)_4(OH)_2]Cl^* + 3(NH_4)_2SO_4$$

*For simplicity a mononuclear complex has been written here even though the chromium(III) ammonia solution contains hydroxo-bridged complexes to a large extent.

$$[Cr(NH_3)_4(OH)_2]Cl + 2HCl \longrightarrow [Cr(NH_3)_4(H_2O)Cl]Cl_2 + H_2O$$

$$[Cr(NH_3)_4(H_2O)Cl]Cl_2 + (NH_4)_2SO_4 \longrightarrow$$
$$[Cr(NH_3)_4(H_2O)Cl]SO_4 + 2NH_4Cl$$

The *cis*-[tetraammineaquachlorochromium(III)] ion has a long history associated in early years with the names Fremy,[6] Cleve,[7] and Jørgensen[8] and later with the name Mori.[9] All these authors dissolved aquachromium(III)[6,9] or chloro complexes[7,8] in concentrated aqueous ammonia with[6-8] or without[9] an added ammonium salt. The present method of preparation is essentially that of Mori[9] but with added ammonium chloride and without heating of the acidic reaction mixture. The corresponding *trans*-isomers were discovered recently.[10,11]

Procedure

Ammonium chloride (32 g, 0.6 mole) is added to aqueous ammonia (150 mL, 25%) and the mixture is heated to 35°, forming a homogeneous solution. Finely powdered chromium(III) sulfate octadecahydrate (72 g, 0.1 mole) is added all at once with vigorous stirring. A temperature increase to about 45° is observed and during ½ hour at this temperature the precipitated basic chromium(III) salt dissolves more or less completely to form a dark-red solution. This solution is then, under cooling, poured (slowly at the beginning) into concentrated hydrochloric acid (200 mL) precooled in ice-water. The temperature should not exceed 40° during this neutralization. After cooling to about 0° the precipitated ammonium chloride containing a small amount of red crystals, mainly pentaamminechlorochromium(III) chloride, is discarded. Concentrated hydrochloric acid (350 mL) is added to the filtrate, which is left overnight in the dark at room temperature. The red crystals that form consist essentially of tetraammineaquachlorochromium(III) chloride. These crystals are washed free of ammonium chloride and green impurities with hydrochloric acid (4 *M*) and are then washed with alcohol and diethyl ether. The yield is 10-16 g.* For purification, tetraammineaquachlorochromium(III) chloride (12 g, 0.05 mole) in a mortar is treated with ice-cold nitric acid (240 mL, 0.1 *M*) for a few minutes and the filtrate† is collected in a flask containing ammonium sulfate (24 g). After shaking and standing at 0° for 2 hours, the bluish-red crystals are collected and washed with ice-water, alcohol, and diethyl ether. The yield of tetraammineaquachlorochrom-

*Another crop of crystals separates from the filtrate during the following days, 2-4 g, which, however, contains an impurity of greenish-blue crystals, probably *fac*-$[Cr(NH_3)_3Cl_3]$. The yield of this substance may be increased by using 700 mL rather than 350 mL of concentrated hydrochloric acid.

†Sometimes retreatment of the nondissolved red crystals with another portion of nitric acid, say 30 mL, and precipitation with ammonium sulfate (3 g) gives a small additional crop of equally pure tetraammineaquachlorochromium(III) sulfate.

ium(III) sulfate is 8 g. The yield based upon chromium(III) sulfate varies between 25 and 40%. *Anal.* Calcd. for $[Cr(NH_3)_4(H_2O)Cl]SO_4$; Cr, 19.29; NH_3, 25.26; Cl, 13.15; SO_4, 35.62; Found: Cr, 18.35; NH_3, 24.64; Cl, 13.57; SO_4, 34.84, corresponding to the ratios $Cr/NH_3/Cl/SO_4$ = 1:4.10:1.08:1.08. Such a sample was used as starting material for the preparation of the aquahydroxochromium complex. The sulfate may be transformed into the corresponding chloride by treatment with hydrochloric acid (6 *M*) in a mortar and a purified sulfate may then be prepared as described above.

Properties

cis-[Tetraammineaquachlorochromium(III)] sulfate forms bluish-red crystals that are sparingly soluble in water. The sulfate is easily transformed into the corresponding chloride (see above), which is quite soluble and from whose aqueous solution the sulfate can be reisolated. The cation hydrolyzes readily in water. The sulfate is used in the synthesis of *cis*-[tetraammineaquahydroxochromium(III)] dithionate (Sec. B).

B. *cis*-[TETRAAMMINEAQUAHYDROXOCHROMIUM(III)] DITHIONATE

$$cis\text{-}[Cr(NH_3)_4(H_2O)Cl]SO_4 + Na_2[S_2O_6] + H_2O + C_5H_5N \longrightarrow$$
$$cis\text{-}[Cr(NH_3)_4(H_2O)(OH)][S_2O_6] + Na_2SO_4 + (C_5H_5NH)Cl$$

cis-[Tetraammineaquahydroxochromium(III)] sulfate is traditionally prepared by the hydrolysis of the corresponding aquachloro complex in a hot-water solution, followed by addition of pyridine and precipitation with ethanol.[12] This method, which requires a great deal of technique, often gives oily products and cannot be recommended for quantities larger than 1 g.

By an analogous method the dithionate salt has been obtained from *cis*-[tetraammineaquahydroxochromium(III)] chloride.[12,13] In the method described here, the aquachloro complex is hydrolyzed at room temperature in a water-pyridine mixture and the resulting *cis*-[tetraammineaquahydroxochromium(III)] complex is isolated as the dithionate salt. The crude product is contaminated with a small amount of sulfate but has a purity suitable for the syntheses of di-μ-hydroxo-bis[tetraamminechromium(III)] salts given in Section H. *cis*-[Tetraammineaquachlorochromium(III)] sulfate is described in Section A.

Procedure

Fifty milliliters of pyridine and 27.0 g (0.1 mole) of *cis*-[tetraammineaquachlorochromium(III)] sulfate is added to a solution of 35.0 g (0.145 mole) of sodium dithionate dihydrate in 325 mL of water at room temperature (20°). The suspension is stirred at room temperature for 35 minutes and then cooled in ice for ½ hour. During this time the bluish-red sulfate salt dissolves and orange crystals of *cis*-[tetraammineaquahydroxochromium(III)] dithionate separate. The sample is filtered and washed with two 20-mL portions of ice-cold water, two 25-mL portions of 50% v/v ethanol-water, and two 25-mL portions of 96% ethanol. Drying in air yields 18 g (54%) of the crude dithionate salt. *Anal.* Calcd. for $[Cr(NH_3)_4(H_2O)(OH)]S_2O_6 \cdot H_2O$: Cr 15.61; N, 16.82; H, 5.44. Found: Cr, 15.39; N, 16.87; H, 5.05.

Properties

cis-[Tetraammineaquahydroxochromium(III)] dithionate is sparingly soluble in water but very soluble in strong acid and strong base, giving solutions of the *cis*-diaqua and *cis*-dihydroxo complex ions, respectively. In the ammonia buffer region fast hydrolysis with loss of ammonia takes place. The crude salt is used in the preparation of *cis*-[tetraamminediaquachromium(III)] perchlorate (Sec. D) and di-μ-hydroxo-bis[tetraamminechromium(III)] dithionate (Sec. H).

C. *cis*-[TETRAAMMINEAQUAHYDROXOCOBALT(III)] DITHIONATE

$$[Co(NH_3)_4(CO_3)]_2SO_4 \cdot 3H_2O + 2H_2SO_4 \longrightarrow$$
$$cis\text{-}[Co(NH_3)_4(H_2O)_2]_2(SO_4)_3 + H_2O + 2CO_2$$

$$cis\text{-}[Co(NH_3)_4(H_2O)_2]_2(SO_4)_3 + 2Na_2[S_2O_6] + 2C_5H_5N \longrightarrow$$
$$2cis\text{-}[Co(NH_3)_4(H_2O)(OH)](S_2O_6) + 2Na_2SO_4 + (C_5H_5NH)_2SO_4$$

In the literature *cis*-[tetraammineaquahydroxocobalt(III)] sulfate[14,15a,b] has been prepared from *cis*-[tetraamminediaquacobalt(III)] sulfate,[16] which is prepared from [tetraammine(carbonato)cobalt(III)] sulfate. In the method described here *cis*-[tetraammineaquahydroxocobalt(III)] dithionate is prepared in this manner from [tetraammine(carbonato)cobalt(III)] sulfate by way of the *cis*-diaqua salt, which is isolated only as a crude product. [Tetraammine(carbonato)-cobalt(III)] sulfate is easily obtained in high yield from cobalt(II) sulfate.[17]

Procedure

To crude [tetraammine(carbonato)cobalt(III)] sulfate trihydrate[17] (21.0 g, 0.080 mole) is added 120 mL of cold (0-5°) 1 *M* sulfuric acid, with stirring and cooling in an ice bath. The carbonato complex dissolves with evolution of carbon dioxide gas and formation of a red solution of the corresponding *cis*-diaqua complex. After 10 minutes the solution is filtered and 60 mL of methanol is added dropwise within a few minutes, with continued cooling and stirring. Microscopic red crystals of *cis*-[tetraamminediaquacobalt(III)] sulfate precipitate. After the suspension has cooled for another 10 minutes, the precipitate is filtered and washed with two 50-mL portions of 96% ethanol and then thoroughly washed with diethyl ether. After it is dried in air the sample (approximately 25 g) is dissolved at room temperature in a solution of 29 g (0.120 mole) of sodium dithionate dihydrate in 400 mL of water. Pyridine (85 mL) is then added carefully to the filtered solution with stirring and cooling in an ice bath. Reddish-purple crystals of *cis*-[tetraammineaquahydroxocobalt(III)] dithionate separate. Cooling is continued for 1½ hours to complete the precipitation. The sample is filtered and washed with two 30-mL portions of water, 30 mL of 50% v/v ethanol-water, and then 96% ethanol. Drying in air yields 18.7 g (69%) of the almost pure dithionate salt. *Anal.* Calcd. for $[Co(NH_3)_4(H_2O)(OH)]S_2O_6 \cdot H_2O$: Co, 17.33; N, 16.47; H, 5.04. Found: Co, 17.34, N, 16.51; H, 4.96.

Properties

cis-[Tetraammineaquahydroxocobalt(III)] dithionate is sparingly soluble in water but dissolves in strong acid or strong base, giving solutions of the corresponding *cis*-diaqua and *cis*-dihydroxo complexes, respectively. The crude salt is used in the preparation of di-μ-hydroxo-bis[tetraamminecobalt(III)] dithionate (Sec. I).

D. *cis*-[TETRAAMMINEDIAQUACHROMIUM(III)] PERCHLORATE

$$cis\text{-}[Cr(NH_3)_4(H_2O)(OH)][S_2O_6] \cdot H_2O + 3HClO_4 \longrightarrow cis\text{-}[Cr(NH_3)_4(H_2O)_2](ClO_4)_3 + H_2S_2O_6 + H_2O$$

cis-[Tetraamminediaquachromium(III)] perchlorate is obtained in the synthesis described here from *cis*-[tetraammineaquahydroxochromium(III)] dithionate (Sec. B) by treatment with perchloric acid. The chloride, bromide, and nitrate salts may be prepared by analogous procedures.[13]

Procedure

■ **Caution.** *Danger! Complex ammines as perchlorate salts may explode. See p. 78. The wash liquids containing perchlorates and organic substances, such as C_2H_5OH, should* not *be heated. The solid perchlorates are dried at room temperature.*

Crude *cis*-[tetraammineaquahydroxochromium(III)] dithionate (6.00 g, 0.018 mole) is dissolved in 20 mL of ice-cold 1.2 *M* perchloric acid (0.024 mole). To the filtered solution is added, with stirring and cooling in an ice bath, 20 mL of ice-cold 70% perchloric acid. Orange crystals of *cis*-[tetraamminediaquachromium(III)] perchlorate precipitate immediately. After the suspension has cooled for 5 minutes, the sample is filtered, washed with 10 mL of 96% ethanol, then dried thoroughly with diethyl ether. Drying in air yields 5.8 g (71%) of the almost pure perchlorate salt. The crude product is purified by dissolving a 4.50-g quantity in 10 mL of ice-cold 0.12 *M* perchloric acid and adding to the filtered solution, with stirring and cooling, 10 mL of ice-cold 70% perchloric acid. The perchlorate salt is isolated as above. The yield is 3.7 g (82%). A further reprecipitation yields a pure product. *Anal.* Calcd. for $[Cr(NH_3)_4(H_2O)_2](ClO_4)_3$: Cr, 11.44; Cl, 23.40; N, 12.33; H, 3.55. Found: Cr, 11.47; Cl, 23.38; N, 12.37; H, 3.47.

Properties

cis-[Tetraamminediaquachromium(III)] perchlorate is stable for months when kept in the cold (-20°). The salt is very soluble in water. The acid dissociation constants (25°, 1 *M* $NaNO_3$) are $pK_1 = 5.08$ and $pK_2 = 7.36$.[18]

E. *cis*-[TETRAAMMINEDIAQUACOBALT(III)] PERCHLORATE

$$[Co(NH_3)_4(CO_3)]ClO_4 + 2HClO_4 + H_2O \longrightarrow$$
$$cis\text{-}[Co(NH_3)_4(H_2O)_2](ClO_4)_3 + CO_2$$

cis-[Tetraamminediaquacobalt(III)] perchlorate has been prepared by addition of perchloric acid to [tetraammine(carbonato)cobalt(III)] perchlorate followed by evaporation of the reaction mixture.[19] In the method given below this reaction is used except that the evaporation of the reaction mixture has been avoided, saving time and avoiding danger. In both procedures the yield is almost quantitative. [Tetraammine(carbonato)cobalt(III)] perchlorate may be obtained from the corresponding sulfate salt.[17,19]

Procedure

■ **Caution.** *Perchlorates may explode. See p. 78.*

[Tetraammine(carbonato)cobalt(III)] perchlorate[17,19] (5.74 g, 0.020 mole) is dissolved in 30 mL of 3 *M* perchloric acid at room temperature with stirring. The carbonato complex dissolves rapidly evolving carbon dioxide. After 5 minutes the solution is filtered and cooled with ice. To the cold (approximately 5°)

solution is then added with stirring 14 mL of ice-cold 70% perchloric acid. Red *cis*-[tetraamminediaquacobalt(III)] perchlorate precipitates almost instantaneously. After the suspension is cooled for another 10 minutes, the sample is filtered, washed with 15 mL of ice-cold 96% ethanol, and washed thoroughly with diethyl ether. Drying in air yields 6.92 g (75%). The sample is reprecipitated as was the chromium(III) compound (see Sec. D) in a yield of 82%. One reprecipitation gives a pure compound. *Anal.* Calcd. for $[Co(NH_3)_4(H_2O)_2]$-$(ClO_4)_3$: Co, 12.77; N, 12.14; Cl, 23.05; H, 3.49. Found: Co, 12.83; N, 12.16; Cl, 23.16; H, 3.47.

Properties

cis-[Tetraamminediaquacobalt(III)] perchlorate is very soluble in water. The acid dissociation constants (20°, 0.1 *M* $NaClO_4$) are $pK_1 = 5.69$ and $pK_2 = 7.99$.[20]

F. *cis*-[AQUABIS(ETHYLENEDIAMINE)HYDROXOCHROMIUM(III)] DITHIONATE

$$cis\text{-}[Cr(en)_2Cl_2]Cl + 2H_2O \longrightarrow cis\text{-}[Cr(en)_2(H_2O)_2]Cl_3$$

$$cis\text{-}[Cr(en)_2(H_2O)_2]Cl_3 + C_5H_5N + Na_2[S_2O_6] \longrightarrow$$
$$cis\text{-}[Cr(en)_2(H_2O)(OH)][S_2O_6] + (C_5N_5HN)Cl + 2NaCl$$

cis-[Aquabis(ethylenediamine)hydroxochromium(III)] dithionate is obtained by hydrolysis of *cis*-[dichlorobis(ethylenediamine)chromium(III)] chloride. The method given here is a slight modification of that given in the literature.[21] The *cis*-dichloro salt is obtained either by thermal decomposition of [tris(ethylenediamine)chromium(III)] chloride[22] or directly from $[Cr(H_2O)_4Cl_2]Cl \cdot 2H_2O$.[23] With the former method the crude chloride salt, and with the latter method, the chloride salt purified by one reprecipitation with hydrochloric acid, are both suitable as starting materials. The crude, almost pure dithionate salt may be dissolved in acid and reprecipitated by addition of base. This procedure, however, does not give a pure product. Pure *cis*-[aquabis(ethylenediamine)hydroxochromium(III)] dithionate is obtained from pure *cis*-[diaquabis(ethylenediamine)-chromium(III)] bromide.

Procedure

cis-[Dichlorobis(ethylenediamine)chromium(III)] chloride monohydrate (20.0 g, 0.0674 mole) is added to 80 mL of hot (60-65°) water, and the mixture is kept at that temperature for ½ hour with stirring. The resulting reddish-purple solution is filtered and cooled in ice. To the cold (5-10°) solution is then added a hot (80-90°) solution of 20 g (0.083 mole) sodium dithionate dihydrate in 80 mL of water, and the mixture is cooled again (to 5-10°). Then 20 mL of pyridine is added with continued stirring, and the mixture is cooled for another hour. Red crystals of *cis*-[aquabis(ethylenediamine)hydroxochromium(III)] dithionate precipitate. The sample is filtered and washed with two 40-mL portions of 50% v/v ethanol-water and two 40-mL portions of 96% ethanol and is then thoroughly dried with diethyl ether. Drying in air yields 17.0 g (69%) of almost pure *cis*-[aquabis(ethylenediamine)hydroxochromium(III)] dithionate. *Anal.* Calcd. for $[Cr(en)_2(H_2O)(OH)][S_2O_6]$: Cr, 14.16; N, 15.25; C, 13.08; H, 5.22. Found: Cr, 14.01; N, 15.07; C, 13.22; H, 5.14.

The pure salt is obtained by the following method from pure *cis*-[diaquabis(ethylenediamine)chromium(III)] bromide dihydrate. *cis*-[Diaquabis(ethylenediamine)chromium(III)] bromide dihydrate (2.00 g, 0.00413 mole) (see Sec. G) is dissolved at 0° in 5.00 mL of 0.100 *M* hydrochloric acid. To this solution is added a solution of 1.10 g (0.00454 mole) of sodium dithionate dihydrate in 10 mL of water at room temperature. To the filtered solution is added, with stirring and cooling in an ice bath, 23.0 mL of 0.200 *M* sodium hydroxide. Precipitation of red crystals of *cis*-[aquabis(ethylenediamine)hydroxochromium(III)] dithionate commences almost instantaneously. After a few minutes the sample is filtered and washed thoroughly with water, 96% ethanol, and then diethyl ether. Drying in air yields 0.9 g (59%) of pure *cis*-[aquabis(ethylenediamine)hydroxochromium(III)] dithionate. *Anal.* Found: Cr, 14.04; N, 15.16; C, 13.17; H, 5.23.

Properties

The *cis*-[aquabis(ethylenediamine)hydroxochromium(III)] dithionate salt is stable for years. It is insoluble in water but very soluble in strong acid and strong base, giving the corresponding *cis*-diaqua and *cis*-dihydroxo species. The crude product is used in the preparation of *cis*-[diaquabis(ethylenediamine)chromium(III)] bromide (Sec. G) and of di-μ-hydroxobis[bis(ethylenediamine)chromium(III)] dithionate (Sec. J).

G. *cis*-[DIAQUABIS(ETHYLENEDIAMINE)CHROMIUM(III)] BROMIDE

$$cis\text{-}[Cr(en)_2(H_2O)(OH)][S_2O_6] + 3HBr + 2H_2O \longrightarrow cis\text{-}[Cr(en)_2(H_2O)_2]Br_3 \cdot 2H_2O + H_2S_2O_6$$

cis-[Diaquabis(ethylenediamine)chromium(III)] bromide was originally prepared by Pfeiffer with [diaquadihydroxobis(pyridine)chromium(III)] chloride as starting material.[24] The preparation given below is based on an improved method.[21] Here the bromide salt is obtained from crude *cis*-[aquabis(ethylenediamine)hydroxochromium(III)] dithionate (Sec. F) by treatment with hydrobromic acid. The pure *cis*-[diaquabis(ethylenediamine)chromium(III)] bromide is used to prepare pure *cis*-[aquabis(ethylenediamine)hydroxochromium(III)] dithionate (Sec. F).

Procedure

Crude *cis*-[aquabis(ethylenediamine)hydroxochromium(III)] dithionate (20.0 g, 0.0544 mole) is dissolved at 0° in 60 mL of 1 *M* hydrobromic acid to give an orange solution of the *cis*-diaqua complex. To the filtered solution is then added, with stirring and cooling in an ice bath, 120 mL of ice-cold 63% hydrobromic acid. Cooling is continued for ½ hour and orange crystals of *cis*-[diaquabis(ethylenediamine)chromium(III)] bromide dihydrate separate. The sample is filtered and washed thoroughly with 96% ethanol. Drying in air yields 17.7 g (67%) of almost pure product. Ten grams of this complex is dissolved in 30 mL of ice-cold 0.01 *M* hydrobromic acid and reprecipitated with 60 mL of 63% hydrobromic acid as above. This yields 6.3 g (63%) of pure *cis*-[diaquabis(ethylenediamine)chromium(III)] bromide dihydrate. *Anal.* Calcd. for $[Cr(en)_2(H_2O)_2]Br_3 \cdot 2H_2O$: Cr, 10.75; N, 11.58; C, 9.93; H, 5.00; Br, 49.54. Found: Cr, 10.70; N, 11.57; C, 9.81; H, 4.91; Br, 49.80.

Properties

The *cis*-[diaquabis(ethylenediamine)chromium(III)] bromide salt is sensitive to light. If stored in the dark at −20°, it is stable for years. In acidified solution it is converted slowly to an equilibrium mixture containing the trans-isomer. The equilibrium constant is K = [cis]/[trans] = 12.8 (25°, 1 *M* $NaNO_3$).[21] The acid dissociation constants are $pK_1 = 4.8$ and $pK_2 = 7.17$ (25°, 1 *M* $NaNO_3$).[21]

H. DI-μ-HYDROXO-BIS[TETRAAMMINECHROMIUM(III)] BROMIDE AND PERCHLORATE

$$2\,cis\text{-}[Cr(NH_3)_4(H_2O)(OH)][S_2O_6]\cdot H_2O \longrightarrow [(NH_3)_4Cr(OH)_2Cr(NH_3)_4][S_2O_6]_2 + 4H_2O$$

$$[(NH_3)_4Cr(OH)_2Cr(NH_3)_4][S_2O_6]_2 + 4NH_4Br + 4H_2O \longrightarrow [(NH_3)_4Cr(OH)_2Cr(NH_3)_4]Br_4\cdot 4H_2O + 2(NH_4)_2[S_2O_6]$$

$$[(NH_3)_4Cr(OH)_2Cr(NH_3)_4]Br_4 \cdot 4H_2O + 4NaClO_4 \longrightarrow$$

$$[(NH_3)_4Cr(OH)_2Cr(NH_3)_4](ClO_4)_4 \cdot 2H_2O + 4NaBr + 2H_2O$$

The sulfate salt of di-μ-hydroxo-bis[tetraamminechromium(III)] has been prepared traditionally by heating of *cis*-[tetraammineaquahydroxochromium(III)] sulfate in acetic anhydride.[12] In the method given here, the impure dithionate salt is obtained by heating *cis*-[tetraammineaquahydroxochromium(III)] dithionate at 100°. The pure bromide salt is then obtained from the crude dithionate, and the pure perchlorate is obtained from the bromide.

Procedure

■ **Caution.** *Perchlorates may explode. See p. 78.*

Crude *cis*-[tetraammineaquahydroxochromium(III)] dithionate (2.00 g, 0.0060 mole) is heated for 1 hour and 15 minutes at 100° in an oven to give a violet product of very impure di-μ-hydroxo-bis[tetraamminechromium(III)] dithionate. This material is added to 8 mL of a saturated (at room temperature) solution of ammonium bromide, and the suspension is cooled in an ice bath, with thorough stirring, for ½ hour. The dithionate salt dissolves and red crystals of di-μ-hydroxo-bis[tetraamminechromium(III)] bromide tetrahydrate precipitate. The sample is filtered, washed with four 5-mL portions of 50% v/v ethanol-water, and dried in the air. This procedure yields 1.00 g (50%) of an almost pure sample. A pure product is obtained after two further reprecipitations. A 1.00-g quantity is dissolved in 10 mL of 0.01 *M* hydrobromic acid and reprecipitated from the filtered solution by addition of 10 mL of the saturated solution of ammonium bromide with stirring and cooling in an ice bath. The bromide salt is isolated as above in a yield of 0.75 g (75%). *Anal.* Calcd. for $[(NH_3)_4Cr(OH)_2Cr(NH_3)_4]Br_4 \cdot 4H_2O$; Cr, 15.62; Br, 48.00; N, 16.83; H, 5.15. Found: Cr, 15.50; Br, 48.16; N, 16.88; H, 4.87.

The perchlorate is prepared from the crude bromide. The bromide (4.0 g, 0.0060 mole) is dissolved in 100 mL of 0.012 *M* perchloric acid at room temperature. To the filtered solution is added, with stirring and cooling in an ice bath, 60 mL of a saturated solution of sodium perchlorate. The mixture is cooled for 1 hour, during which time violet crystals of di-μ-hydroxo-bis[tetraamminechromium(III)] perchlorate dihydrate separate. The sample is filtered, washed with three 10-mL portions of 96% ethanol, and dried in air. This procedure yields 3.9 g (92%) of a pure product. A 2.00-g quantity is reprecipitated by dissolution in 25 mL of 0.012 *M* perchloric acid at room temperature and addition of 10 mL of a saturated solution of sodium perchlorate as above. This method yields 1.9 g (95%). *Anal.* Calcd. for $[(NH_3)_4Cr(OH)_2Cr(NH_3)_4](ClO_4)_4 \cdot 2H_2O$: Cr, 14.69; Cl, 20.02; N, 15.83; H, 4.27. Found: Cr, 14.62; Cl, 20.00; N, 15.72; H, 4.06.

Properties

See Section K.

I. DI-μ-HYDROXO-BIS[TETRAAMMINECOBALT(III)] BROMIDE AND PERCHLORATE

$$2\ cis\text{-}[Co(NH_3)_4(H_2O)(OH)][S_2O_6] \xrightarrow{\text{heat}} [(NH_3)_4Co(OH)_2Co(NH_3)_4][S_2O_6]_2 + 2H_2O$$

$$[(NH_3)_4Co(OH)_2Co(NH_3)_4][S_2O_6]_2 + 4NH_4Br + 4H_2O \longrightarrow [(NH_3)_4Co(OH)_2Co(NH_3)_4]Br_4 \cdot 4H_2O + 2(NH_4)_2[S_2O_6]$$

$$[(NH_3)_4Co(OH)_2Co(NH_3)_4]Br_4 \cdot 4H_2O + 4NaClO_4 \longrightarrow [(NH_3)_4Co(OH)_2Co(NH_3)_4](ClO_4)_4 \cdot 2H_2O + 4NaBr + 2H_2O$$

The sulfate salt of di-μ-hydroxo-bis[tetraamminecobalt(III)] has been obtained traditionally by heating *cis*-[tetraammineaquahydroxocobalt(III)] sulfate.[12,15b] In the method given below the crude dithionate salt is obtained almost quantitatively by heating *cis*-[tetraammineaquahydroxocobalt(III)] dithionate at 110°. The pure bromide salt is then obtained from the crude dithionate, and the pure perchlorate is obtained from the bromide.

Procedure

■ **Caution.** *Perchlorates may explode. See p. 78.*

Crude *cis*-[tetraammineaquahydroxocobalt(III)] dithionate (10.0 g, 0.0294 mole) is heated for 2½ hours at 110° in an oven to give 9.0 g of red violet, crude di-μ-hydroxo-bis[tetraamminecobalt(III)] dithionate. The loss in weight is 10%, compared with the theoretical value 10.6%. Continued heating for another 30 minutes does not give any further loss in weight.

The crude dithionate is added to 35 mL of saturated solution of ammonium bromide and the suspension is cooled for 1 hour in an ice bath with thorough stirring. The dithionate dissolves and violet crystals of di-μ-hydroxo-bis[tetraamminecobalt(III)] bromide separate. The sample is filtered, washed with three 20-mL portions of ice-cold 50% v/v ethanol-water, and dried in air. An almost quantitative yield of the bromide is obtained. This is dissolved in 450 mL of 0.01 *M* hydrobromic acid at room temperature. To the filtered solution is added, with stirring and cooling in an ice bath, 90 mL of a saturated solution of ammonium bromide. Violet crystals of the bromide salt separate almost instantaneously. After cooling the suspension for ½ hour, the sample is filtered and washed with four 20-mL portions of ice cold 50% v/v ethanol-water. Drying in air yields 7.9 g (79%) of the crude bromide salt. The pure salt is obtained by one more reprecipitation. A 1.00-g quantity is dissolved in 50 mL of 0.01 *M* hydrobromic acid at room temperature and reprecipitated as above by addition of 5 mL of a saturated solution of ammonium bromide. Yield, 0.9 g (90%). *Anal.* Calcd. for $[(NH_3)_4Co(OH)_2Co(NH_3)_4]Br_4 \cdot 4H_2O$: Co, 17.34; Br, 47.02; N, 16.49; H, 5.05. Found: Co, 17.37; Br, 47.28; N, 16.57; H, 4.81.

The perchlorate salt is prepared from the crude bromide. A 4.00-g (0.059 mole) quantity of crude bromide is added to 30 mL of a saturated solution of sodium perchlorate and the suspension is kept at room temperature, with stirring, for 2 hours. During this time the bromide salt dissolves and red crystals of di-μ-hydroxo-bis[tetraamminecobalt(III)] perchlorate dihydrate separate almost quantitatively. The sample is filtered, washed with three 10-mL portions of 96% ethanol, and dried in air. This sample is then dissolved in 70 mL of 0.012 *M* perchloric acid at room temperature. To the filtered solution is then added, with stirring and cooling in an ice bath, 20 mL of a saturated solution of sodium perchlorate. After cooling for 1 hour the sample is isolated as above. This yields 3.8 g (89%) of the pure perchlorate. Reprecipitation of 2.0 g by dissolution in 35 mL of 0.012 *M* perchloric acid and addition of 10 mL of a saturated solution of sodium perchlorate yields 1.7 g (85%). *Anal.* Calcd. for $[(NH_3)_4Co(OH)_2Co(NH_3)_4](ClO_4)_4 \cdot 2H_2O$: Co, 16.33; N, 15.53; H, 4.20; Cl, 19.64. Found: Co, 16.34; N, 15.53; H, 4.06; Cl, 19.60.

Properties

See Section K.

J. DI-μ-HYDROXO-BIS[BIS(ETHYLENEDIAMINE)CHROMIUM(III)] DITHIONATE, BROMIDE, CHLORIDE, AND PERCHLORATE

$$2cis\text{-}[Cr(en)_2(H_2O)(OH)][S_2O_6] \xrightarrow{\text{heat}} [(en)_2Cr(OH)_2Cr(en)_2][S_2O_6]_2 + 2H_2O$$

$$[(en)_2Cr(OH)_2Cr(en)_2][S_2O_6]_2 + 4NH_4Br + 2H_2O \longrightarrow [(en)_2Cr(OH)_2Cr(en)_2]Br_4 \cdot 2H_2O + 2(NH_4)_2[S_2O_6]$$

Di-μ-hydroxo-bis[bis(ethylenediamine)chromium(III)] salts are traditionally obtained from the dithionate salt. The dithionate may be prepared by heating of *cis*-[aquabis(ethylenediamine)hydroxochromium(III)] dithionate either at 100-120° or by refluxing in acetic anhydride.[12] Alternatively, hydrolysis in a water-pyridine mixture of *cis*-[diaquabis(ethylenediamine)chromium(III)] bromide yields the bromide salt of the bridged cation in low yield.[25]

In the procedure given here the dithionate is obtained almost quantitatively from *cis*-[aquabis(ethylenediamine)hydroxochromium(III)] dithionate by refluxing in acetic anhydride, following the method given in the literature.[12] The preparation of *cis*-[aquabis(ethylenediamine)hydroxochromium(III)] dithionate is given in Section F.

The bromide and chloride salts have been obtained in high yields from the dithionate by modifications of the procedures given in the literature.[12] The perchlorate salt is obtained from the bromide. Other salts, such as the iodide, thiocyanate, and nitrate, have been described in the literature.[12]

Procedure

■ **Caution.** *Perchlorates may explode. See p. 78.*

Crude *cis*-[aquabis(ethylenediamine)hydroxochromium(III)] dithionate (40.0 g, 0.109 mole) (see Sec. E) is added to 400 mL of acetic anhydride. The suspension is heated to reflux within 20 minutes, kept at reflux for another 20 minutes, and then cooled in an ice bath. The sample is filtered, washed with two 50-mL portions of 96% ethanol, five 50-mL portions of 2 *M* acetic acid, and four 100-mL portions of 96% ethanol, and then thoroughly washed with diethyl ether. By the washing with acetic acid, a small amount of unreacted *cis*-[aquabis(ethylenediamine)hydroxochromium(III)] dithionate is removed.* Drying in air yields 34.5 g (91%) of a nearly pure product. The crude dithionate is used in the syntheses of the chloride and the bromide, as given below. The pure dithionate is obtained from the pure bromide by the following procedure.

Pure di-μ-hydroxo-bis[bis(ethylenediamine)chromium(III)] bromide (0.50 g, 0.00068 mole) is dissolved in 100 mL of 0.01 *M* hydrobromic acid at 5°. To the fitered solution is added 10 mL of a saturated solution of sodium dithionate dihydrate and the solution is kept at 5° for 24 hours. Large red crystals of the dithionate separate. The sample is filtered, thoroughly washed with water, and dried in air. This yields 0.2 g (42%). *Anal.* Calcd. for $[(en)_2Cr(OH)_2Cr(en)_2][S_2O_6]_2$: Cr, 14.89; C, 13.75; N, 16.04; H, 4.91. Found: Cr, 14.77; C, 13.75; N, 15.92; H, 4.95.

The crude dithionate salt (10.0 g, 0.0143 mole) is added to 25 mL of a saturated solution of ammonium bromide and the suspension is kept at room temperature with stirring for 1 hour. The violet crystals of the bromide salt are filtered and washed with two 25-mL portions of 50% v/v ethanol-water. The product is

*It should be noted that mixing of the mother liquor with ethanol is not advisable because of the possibility of the vigorous formation of ethyl acetate.

sucked as dry as possible and then dissolved in 1250 mL of 0.01 *M* hydrobromic acid at room temperature. Then 100 mL of a saturated solution of ammonium bromide is added portionwise over a period of 10 minutes to the filtered solution, which is then cooled slowly in an ice bath. After it is cooled for 1 hour the sample is filtered, washed with two 25-mL portions of 50% v/v ethanol-water and two 25-mL portions of 96% ethanol, and dried over 5.4 *M* sulfuric acid. This yields 8.65 g (82%) of nearly pure di-μ-hydroxo-bis[bis(ethylenediamine)chromium(III)] bromide dihydrate. The complex (5.00 g) is dissolved in 625 mL of 0.01 *M* hydrobromic acid and reprecipitated with 50 mL of a saturated solution of ammonium bromide, as above. The yield is 4.80 g (96%). This product was reprecipitated once more to give a pure salt. *Anal.* Calcd. for $[(en)_2Cr(OH)_2Cr(en)_2]Br_4 \cdot 2H_2O$: Cr, 14.17; C, 13.09; H, 5.22; N, 15.27; Br, 43.54. Found: Cr, 14.15; C, 13.08; H, 5.26; N, 15.56; Br, 43.44.

The crude dithionate (10.0 g, 0.0143 mole) is added to 50 mL of a saturated solution of ammonium chloride and the suspension is stirred at room temperature for ½ hour. Violet crystals of the chloride salt are filtered, washed with two 20-mL portions of 50% v/v ethanol-water, and dried in air. The sample is dissolved in 130 mL of 0.01 *M* hydrochloric acid at room temperature and filtered. Then 160 mL of a saturated solution of ammonium chloride is added portionwise during 10 minutes to the stirred solution, which is then cooled slowly in an ice bath. After it is cooled for 1 hour, the sample if filtered, washed with 15 mL of 50% v/v ethanol-water and two 15-mL portions of 96% v/v ethanol-water and dried over 5.4 *M* sulfuric acid. This yields 5.0 g (63%) of the almost pure di-μ-hydroxo-bis[bis(ethylenediamine)chromium(III)] chloride dihydrate. A 2.00-g, quantity is dissolved in 35 mL of 0.01 *M* hydrochloric acid and reprecipitated with 40 mL of a saturated solution of ammonium chloride, as above. The yield is 1.55 g (78%) of pure salt. *Anal.* Calcd. for $[(en)_2Cr(OH)_2Cr(en)_2]Cl_4 \cdot 2H_2O$: Cr, 18.70; C, 17.27; N, 20.15; H, 6.89; Cl, 25.49. Found: Cr, 18.46; C, 17.28; N, 20.13; H, 6.75; Cl, 25.56.

The perchlorate is obtained from the crude bromide. The crude bromide (3.00 g, 0.0041 mole) is added to a mixture to 20 mL of a saturated solution of sodium perchlorate and 20 mL of water, and the suspension is stirred at room temperature for 1 hour. The violet crystals of the perchlorate salt are collected on a filter and the sample is sucked as dry as possible and then dissolved in 40 mL of 0.012 *M* perchloric acid. Then 40 mL of a saturated solution of sodium perchlorate is added portionwise over a 10-minute period to the stirred solution, which is thereafter cooled slowly in an ice bath. After 20 minutes of cooling the sample is filtered and washed with 5 mL 50% v/v ethanol-water and two 10-mL portions of 96% ethanol. Drying over concentrated sulfuric acid yields 2.78 g (88%) of pure di-μ-hydroxo-bis[bis(ethylenediamine)chromium(III)] perchlorate. *Anal.* Calcd. for $[(en)_2Cr(OH)_2Cr(en)_2](ClO_4)_4$: Cr, 13.40; C, 12.38; N, 14.44; H, 4.42; Cl, 18.27. Found: Cr, 13.30; C, 12.25; N, 14.34; H, 4.42; Cl, 18.26.

Properties

See Section K.

K. DI-μ-HYDROXO-BIS[BIS(ETHYLENEDIAMINE)COBALT(III)] DITHIONATE, BROMIDE, CHLORIDE, AND PERCHLORATE

$$2cis\text{-}[Co(en)_2(H_2O)(OH)]\,[S_2O_6] \xrightarrow{\text{heat}} [(en)_2Co(OH)_2Co(en)_2]\,[S_2O_6]_2 + 2H_2O$$

$$[(en)_2Co(OH)_2Co(en)_2]\,[S_2O_6]_2 + 4NH_4Br + 2H_2O \longrightarrow [(en)_2Co(OH)_2Co(en)_2]\,Br_4\cdot 2H_2O + 2(NH_4)_2[S_2O_6]$$

Di-μ-hydroxo-bis[bis(ethylenediamine)cobalt(III)] dithionate may be obtained analogously to the chromium(III) salt from *cis*-[aquabis(ethylenediamine)hydroxocobalt(III)] dithionate, either by heating at 110°[26] or by refluxing in acetic anhydride.[12] The formation of the bridged cation of cobalt(III) is much slower than that of chromium(III). In contrast to the chromium(III) complex there is no evidence that the bridged cobalt(III) complex can be formed by aqueous hydrolysis.

In the procedure given below, di-μ-hydroxo-bis[bis(ethylenediamine)cobalt(III)] dithionate is obtained quantitatively from *cis*-[aquabis(ethylenediamine)hydroxocobalt(III)] dithionate by refluxing in acetic anhydride. The *cis*-[aquabis(ethylenediamine)hydroxocobalt(III)] dithionate is obtained from carbonatobis(ethylenediamine)cobalt(III) chloride.[27]

The bromide and the chloride are obtained from the dithionate by modifications of the procedures given in the literature.[12,26] The perchlorate is obtained from the bromide. Other salts, such as the iodide, thiocyanate and nitrate, have been described in the literature.[12,26]

Procedure

■ **Caution.** *Perchlorates may explode. See p. 78.*

Crude *cis*-[aquabis(ethylenediamine)hydroxocobalt(III)] dithionate (40.0 g, 0.107 mole) is added to 400 mL of acetic anhydride. The suspension is refluxed for 3 hours. The crude di-μ-hydroxo-bis[bis(ethylenediamine)cobalt(III)] dithionate is isolated analogously to the chromium(III) complex as mentioned in Section 15-J. Yield is 34.0 g (89%) of a crude, almost pure product.

The crude dithionate salt is used in the synthesis of the chloride and the bromide salts given below. The pure dithionate salt is obtained from the pure bromide.

Pure di-μ-hydroxo-bis[bis(ethylenediamine)cobalt(III)] bromide (2.1 g, 0.0028 mole) is dissolved in 800 mL of 0.01 *M* hydrobromic acid at room temperature. To the filtered solution is added 100 mL of 0.2 *M* sodium dithionate with stirring. After about 10 minutes the sample is filtered, thoroughly washed with water, and dried in air, giving 2.0 g (100%) of the dithionate salt. *Anal.* Calcd. for $[(en)_2Co(OH)_2Co(en)_2][S_2O_6]_2$: Co, 16.54; C, 13.48; N, 15.73; H, 4.81. Found: Co, 16.57; C, 13.46; N, 15.63; H, 4.83. The crude dithionate (10.0 g, 0.0141 mole) is added to 25 mL of a saturated solution of ammonium bromide and the suspension is kept at room temperature with stirring for 1 hour. The violet crystals of the bromide are filtered and washed with two 25-mL portions of 50% v/v ethanol-water. The product is sucked as dry as possible and then dissolved in 200 mL of 0.01 *M* hydrobromic acid at room temperature. To the filtered solution 50 mL of a saturated solution of ammonium bromide is added, with stirring and cooling in an ice bath. After it is cooled for 15 minutes, the sample is filtered, washed with two 25-mL portions of 50% v/v ethanol-water, and dried over 5.4 *M* sulfuric acid. This yields 7.60 g (72%) of almost pure di-μ-hydroxo-bis[bis(ethylenediamine)cobalt(III)] bromide dihydrate. The pure sample is obtained by dissolution of the product (5.00 g) in 150 mL of 0.01 *M* hydrobromic acid and reprecipitation with 25 mL of a saturated solution of ammonium bromide as above. The yield is 4.1 g (82%). *Anal.* Calcd. for $[(en)_2Co(OH)_2Co(en)_2]Br_4 \cdot 2H_2O$: Co, 15.76; C, 12.85; H, 5.12; N, 14.98; Br, 42.74. Found: Co, 15.72; C, 12.90; H, 5.10; N, 15.01; Br, 42.58.

The crude dithionate (10.0 g, 0.0141 mole) is added to 40 mL of a saturated solution of ammonium chloride and the suspension is stirred at room temperature for 1 hour. Red crystals of the chloride salt separate. The sample is filtered, washed with three 7-mL portions of ice-cold 50% v/v ethanol-water, and sucked as dry as possible. The sample is dissolved in 75 mL of 0.005 *M* hydrochloric acid at room temperature and filtered. Then 45 mL of a saturated solution of ammonium chloride is added with stirring and cooling in an ice bath. After cooling for 1 hour, the sample is filtered, washed with three 5-mL portions of 50% v/v ethanol-water, and dried over 5.4 *M* sulfuric acid. This yields 5.4 g (62%) of the crude, almost pure di-μ-hydroxo-bis[bis(ethylenediamine)cobalt(III)] chloride pentahydrate. The pure salt is obtained by dissolution of 2.00 g of the crude chloride salt in 20 mL of 0.005 *M* hydrochloric acid and reprecipitation by addition of 12 mL of a saturated solution of ammonium chloride as above. Yield 1.75 g (88%). *Anal.* Calcd. for $[(en)_2Co(OH)_2Co(en)_2]Cl_4 \cdot 5H_2O$: Co, 18.89; C, 15.39; N, 17.96; H, 7.11; Cl, 22.72. Found: Co, 19.01; C, 15.39; N, 17.90; H, 6.71; Cl, 22.85.

The perchlorate is obtained from the crude bromide. The crude bromide salt (3.00 g, 0.00401 mole) is added to a mixture of 20 mL of a saturated solution of sodium perchlorate and 20 mL of water, and the suspension is stirred for 1 hour at room temperature. The sample is then filtered, sucked as dry as possible, and dissolved in 30 mL of 0.012 *M* perchloric acid at room temperature. The solution is filtered and 30 mL of a saturated solution of sodium perchlorate is added with stirring and cooling in an ice bath. After it is cooled for 20 minutes, the sample is filtered and washed with 3 mL of 50% v/v ethanol-water and two 10-mL portions of 96% ethanol. Drying over concentrated sulfuric acid yields 2.9 g (92%) of a pure product. *Anal.* Calcd. for $[(en)_2Co(OH)_2Co(en)_2](ClO_4)_4$: Co, 14.92; C, 12.16; N, 14.19; H, 4.34; Cl, 17.95. Found: Co, 14.86, C, 11.98; N, 14.28; H, 4.35; Cl, 17.87.

Properties

The di-μ-hydroxo complexes obtained in Sections H-K have several chemical and physical properties in common. The salts are stable for years when pure. However, salts of the $[(NH_3)_4Cr(OH)_2Cr(NH_3)_4]^{4+}$ ion may decompose rapidly when not pure. The dithionate salts are insoluble in water. The bromide, chloride, and perchlorate salts are all very soluble in water, except for $[(en)_2Cr(OH)_2Cr(en)_2]$-$Br_4 \cdot 2H_2O$, which is only moderately soluble.

The dinuclear structure has been established by x-ray analyses of $[(NH_3)_4Co$-$(OH)_2Co(NH_3)_4]Cl_4 \cdot 4H_2O$[28,29] and *meso*-$[(en)_2Cr(OH)_2Cr(en)_2]Cl_2(ClO_4)_2 \cdot$-$2H_2O$.[30] The Guinier x-ray powder diffraction patterns of $[(NH_3)_4Co(OH)_2$-$Co(NH_3)_4]Br_4 \cdot 2H_2O$ and the corresponding chromium(III) salt are almost identical, indicating isomorphism and allowing the conclusion that the cations have the same structure. The corresponding perchlorate salts have dissimilar powder patterns. The powder pattern of *meso*-$[(en)_2Cr(OH)_2Cr(en)_2]Br_4 \cdot 2H_2O$ is almost identical with that of the corresponding cobalt(III) salt, whose cation therefore also is of the *meso*-di-μ-hydroxo type. A similar identity is also found between the two dithionate salts. However, the two perchlorate salts and the two chloride salts, respectively, have dissimilar powder patterns.

The inertness of the dinuclear complexes is greatest in slightly acidic solutions, which therefore have been employed for the reprecipitation reactions. Apparently the chromium systems are much more labile toward bridge breaking than are the cobalt systems. In aqueous solution the *meso*-$[(en)_2Cr(OH)_2Cr(en)_2]^{4+}$ cation (I) enters into a rapidly established ($t_{1/2} \sim 1$ min. at room temperature) equilibrium with the mono-μ-hydroxo complex $[(OH)(en)_2Cr(OH)Cr(en)_2$-$(H_2O)]^{4+}$ (II).[1,2] The equilibrium constant $K = [II]/[I]$ is 0.83 in 1 *M* $NaClO_4$ at 0°. The salts (dithionate, bromide, chloride, and perchlorate) of the di-μ-hydroxo cation are less soluble than the respective salts of the mono-μ-hydroxo cation. It is therefore possible to precipitate the pure salts of the di-μ-hydroxo cation from the equilibrium mixture following the procedure given above.

The data given in the table are based on spectra taken 10 minutes after the time of dissolution (at 20.0°) and therefore represent the spectrum of the equilibrium mixture. The spectrum of the $[(en)_2Cr(OH)_2Cr(en)_2]^{4+}$ cation has been obtained from absorption curves (extrapolated back to $t = 0$) of the perchlorate salt in 1 *M* $NaClO_4$ at 0° and shows $(\epsilon,\lambda)_{max}$ = (199,539.5) and (107,386) and $(\epsilon, \lambda)_{sh}$ = (54,284).[2] Almost identical (ϵ, λ) values were obtained for the first absorption band of the perchlorate salt (198,539) and the chloride (198,539.5) in pure water at 0°. The spectrum of the bromide at 0° was not measured because of the very slow rate of dissolution of this salt. The perchlorate has been used for the preparation of salts of the monohydroxo-bridged complexes.[1,2] In basic solution these salts dissolve with a blue color because of deprotonation of one of the hydroxo bridges (p$K \simeq 12$ in water at 20°).[1] In strongly acidic solutions, hydrolysis to form monomeric species occurs. Thus the $[(en)_2Cr(OH)_2Cr(en)_2]Br_4$ salt gives *cis*-$[Cr(en)_2Cl_2]Cl \cdot H_2O$ with concentrated hydrochloric acid, and *cis*-$[Cr(en)_2(H_2O)(Br)]Br_2$[25] with concentrated hydrobromic acid. The rapid ring-opening and ring-closure, as just discussed for the ethylenediamine complex, have not been investigated for the $[(NH_3)_4Cr(OH)_2Cr(NH_3)_4]^{4+}$ cation. However, when sodium hydroxide is added to an aqueous solution of this complex, the color shifts instantaneously from reddish-purple to blue, probably owing to a deprotonation of one hydroxo bridge, and thereafter rapidly ($t_{1/2} < 1$ sec at room temperature) becomes brownish owing to loss of ammonia. In strongly acid or strongly basic solutions, the Co(III) dinuclear complexes hydrolyze rapidly to give mononuclear compounds as the first isolable products. The $[(NH_3)_4Co(OH)_2Co(NH_3)_4]Cl_4$ salt gives a mixture of *cis*-$[Co(NH_3)_4Cl_2]Cl$ and *cis*-$[Co(NH_3)_4(H_2O)_2]Cl_3$[15b,31,32] on treatment with hydrochloric acid saturated with hydrogen chloride at −12°. With concentrated cold, aqueous ammonia *cis*-$[Co(NH_3)_4(H_2O)(OH)]Cl_2 \cdot H_2O$[12] is obtained. The $[(en)_2Co(OH)_2Co(en)_2]Cl_4 \cdot 2H_2O$ salt gives *cis*-$[Co(en)_2Br_2]Br$[26] on treatment with saturated hydrobromic acid (12°). The hydrolysis of the $[(NH_3)_4Co(OH)_2Co(NH_3)_4]^{4+}$ ion in dilute acid[33] and of the $[(en)_2Co(OH)_2Co(en)_2]^{4+}$ ion in dilute acid and dilute base have been investigated.[34-36] A single bridged complex was assumed to be an intermediate in these reactions. In both the acid and the base hydrolysis the corresponding mononuclear diaqua and dihydroxo species, respectively, were the final products. The acid dissociation constant of the $[(en)_2Co(OH)_2Co(en)_2]^{4+}$ ion has been estimated to be in the range 10^{-10} to 10^{-11} *M*.

References

1. J. Springborg and H. Toftlund, *Chem. Commun.*, **1975**, 422.
2. J. Springborg and H. Toftlund, *Acta Chem. Scand.*, **30**, 171, (1976).
3. J. D. Ellis, K. L. Scott, R. K. Wharton, and A. G. Sykes, *Inorg. Chem.*, **11**, 2565 (1972).
4. J. Springborg and C. E. Schäffer, *Inorg. Chem.*, **15**, 1744 (1976).
5. J. Springborg and C. E. Schäffer, *Acta Chem. Scand.*, **A 30**, 787 (1976).
6. E. Fremy, *Compt. Rend.*, **47**, 883 (1858).
7. P. T. Cleve, *K. Sven. Vetensk. Handl.*, [2] **6** Nr. 4 1, (1865).
8. S. M. Jørgensen, *J. Prakt. Chem.*, **20**, (2) 105 (1879); ibid. **42**, (2) 206 (1890).
9. M. Mori, *J. Inst. Polytech. Osaka City Univ.*, C **3**, 41 (1952).

10. J. Glerup and C. E. Schäffer, *Chem. Commun.*, **1968**, 38.
11. J. Glerup and C. E. Schäffer, *Inorg. Chem.*, **15**, 1408 (1976).
12. J. V. Dubsky, *J. Prakt. Chem.*, **90**, 61 (1914).
13. P. Pfeiffer, *Ber. Chem.*, **40**, 3126 (1907).
14. S. M. Jørgensen, *Z. Anorg. Allgem. Chem.*, **16**, 184 (1898).
15. A. Werner, *Ber. Chem.*, **40**, 4116[a], 4820[b] (1907).
16. S. M. Jørgensen, *Z. Anorg. Allgem. Chem.*, **2**, 281 (1892).
17. G. Schlessinger, *Inorg. Synth.*, **6**, 173 (1960).
18. J. Bjerrum, G. Schwarzenbach, and L. G. Sillén, *Stability Constants*, **2**, 8 (1958) Chemical Society, London.
19. M. Lindhard and M. Weigel, *Z. Anorg. Allgem. Chem.*, **260**, 65 (1949).
20. G. Schwarzenbach, J. Boesch, and H. Egli, *J. Inorg. Nucl. Chem.*, **33**, 2141 (1971).
21. F. Woldbye, *Acta Chem. Scand.*, **12**, 1079 (1958).
22. C. L. Rollinson and J. C. Bailar, Jr., *Inorg. Synth.*, **2**, 200 (1946).
23. E. Pedersen, *Acta Chem. Scand.*, **24**, 3362 (1970).
24. P. Pfeiffer, *Chem. Ber.*, **37**, 4275 (1904).
25. P. Pfeiffer and R. Stern, *Z. Anorg. Allgem. Chem.*, **58**, 272 (1908).
26. A. Werner and J. Rapiport, *Lieb. Ann. Chem.*, **375**, 84 (1910).
27. J. Springborg and C. E. Schäffer, *Inorg. Synth.*, **14**, 63 (1973).
28. C. K. Prout, *J. Chem. Soc.*, **1962**, 4429.
29. N. G. Vannerberg, *Acta Chem. Scand.*, **17**, 85 (1963).
30. K. Kaas, *Acta Cryst.* **B32**, 2021 (1976).
31. A. Werner, *Chem. Ber.*, **41**, 3884 (1908).
32. A. Werner, *Ann. Chem.*, **386**, 16 (1912).
33. A. B. Hoffmann and H. Taube, *Inorg. Chem.*, **7**, 1903 (1968).
34. S. E. Rasmussen and J. Bjerrum, *Acta Chem. Scand.*, **9**, 735 (1955).
35. A. A. El-Awady and Z. Z Hugus, Jr., *Inorg. Chem.*, **10**, 1415 (1971).
36. M. M. de Maine and J. B. Hunt, *Inorg. Chem.*, **10**, 2106 (1971).

16. THE RESOLUTION OF BIS(ETHYLENEDIAMINE)-OXALATOCOBALT(III) ION AND ITS USE AS A CATIONIC RESOLVING AGENT

Submitted by WILLIAM T. JORDAN* and LARRY R. FROEBE†
Checked by ROLAND A. HAINES‡ and T. D. LEAH‡

An excellent resolving agent for many dissymmetric anionic metal complexes[1-6] is $[Co(C_2O_4)(en)_2]^+$.§ This complex ion is prepared easily[7] from inexpensive materials and the optical isomers are relatively stable to isomerization in aqueous solutions. Previously, the optical isomers of $[Co(C_2O_4)(en)_2]^+$ were separated[7] by means of diastereoisomer formation with $[Co(edta)]^-$.§ This

*Department of Chemistry, Pacific University, Forest Grove, OR 97116.
†Regional Air-Pollution Control Agency, 451 W. Third Street, Dayton, OH 45402.
‡Department of Chemistry, University of Western Ontario, London, Canada.

required two preliminary syntheses and resolutions; the $[Co(edta)]^-$ was resolved[8] using *cis*-$[Co(en)_2(NO_2)_2]^+$, which in turn was resolved using $(+)_{589}$-bis [μ-tartrato(4−)]-diantimonate(2−), commonly called the antimonyl tartrate ion.[9] Now a simple, one-step resolution, using hydrogen $(+)_{589}$-tartrate ion, has been formulated[1] that provides both optical isomers (enantiomers) of $[Co(C_2O_4)(en)_2]^+$ in high yield (about 70% each) and in a short period of time (1-2 hr). This resolution procedure works well for many cationic complexes.

To demonstrate the utility of $[Co(C_2O_4)(en)_2]^+$ as a cationic resolving agent, two illustrative resolutions are given here. The resolution of $[Co(edta)]^{-}$[§] yields an anionic resolving agent of considerable usefulness[4,5] in its own right. The method given here[1] provides an attractive alternative to the use of *cis*-$[Co(en)_2(NO_2)_2]^+$.[9] The method[2] for resolving *sym-cis*-$[Co(edda)(NO_2)_2]^-$ is typical of the procedures that have been used successully to resolve a number of anionic complexes containing edda.[2-4]

When the two diastereomers (diastereoisomers) differ significantly in solubility, one can use a 1:1 molar ratio of complex and resolving agent.[10] In this way both enantiomers can be recovered. The more soluble diastereomer commonly needs further purification (Sec. B and references). It might be recrystallized or, in some cases, converted to the enantiomer for purification. The isolation of an optically pure enantiomer from an incompletely resolved complex often depends on different solubilities of the enantiomer and the racemate. Such differences do not occur always. When the solubilities of the two diastereomers do not differ greatly, the more insoluble one can often be separated well using ½ mole of resolving agent per mole of complex (assuming charges of equal magnitude). The second enantiomer might be recovered from the filtrate, but often its optical purity (degree of resolution) is not good. Resolution procedure E yields only one of the optically pure enantiomers, but avoids lengthy and tedious separations.

A. BIS(ETHYLENEDIAMINE)OXALATOCOBALT(III) CHLORIDE MONOHYDRATE, $[Co(C_2O_4)(en)_2]Cl \cdot H_2O$

$$2HCl + 2Co(CH_3CO_2)_2 \cdot 4H_2O + PbO_2 + 4en + 2H_2C_2O_4 + H_2O \longrightarrow$$
$$2[Co(C_2O_4)(en)_2]Cl \cdot H_2O + PbO + 4CH_3CO_2H + 8H_2O$$

§ Abbreviations used for ligands: en = ethylenediamine; edta = anion of ethylenediaminetetraacetic acid [(ethylenedinitrilo)tetraacetic acid]; edda = anion of ethylenediamine-*N,N'*-diacetic acid [(ethylenediimino)diacetic acid; *N,N'*-ethylenediglycine]. Lower case letters are used for edta and edda as recommended by the I.U.P.A.C. nomenclature rules, even though this is not current practice in the literature.

Procedure

A solution of 20 g (0.080 mole) of cobalt(II) acetate tetrahydrate in 100 mL of water at 60° is added to a mixture of 15 g (0.12 mole) of oxalic acid dihydrate and 15 mL (0.22 mole) of 99% ethylenediamine (or an equimolar amount of $en{\cdot}H_2O$) in 100 mL of water at 70°. The reaction mixture is heated rapidly to 80° with continuous stirring and 10 g (0.042 mole) of lead(IV) oxide is added. The resulting mixture is boiled gently for 30 minutes, with an additional 2 g of PbO_2 added after 10 minutes and another 2 g of PbO_2 after 20 minutes. After it is cooled to room temperature and treated with 10 mL of 10 *N* sulfuric acid, the mixture is filtered. Twenty-five milliliters of 10 *N* hydrochloric acid is added to the filtrate, which is then evaporated to 100 mL on a steam bath using an air stream. The solution is cooled in ice and filtered to obtain red crystals, which are washed with 80% methanol, methanol, and diethyl ether, and dried on the funnel by suction. The yield is 21 g (86%). Recrystallization is achieved by dissolution in 150 mL of hot water and addition of 25 mL of concentrated HCl, followed by cooling in ice. The crystals are filtered and washed as before. *Anal.* Calcd. for $[Co(C_2H_8N_2)_2C_2O_4]Cl{\cdot}H_2O$: C, 22.48; H, 5.66; N, 17.48. Found: C, 22.70; H, 5.76; N, 17.69.

Properties

The product can be characterized by two absorption bands, $\epsilon_{500} = 103$ and $\epsilon_{360} = 150$.

B. RESOLUTION OF BIS(ETHYLENEDIAMINE)OXALATOCOBALTATE-(III) ION

$$2[Co(C_2O_4)(en)_2]Cl{\cdot}H_2O + (+)_{589}\text{-}H_2C_4H_4O_6 + Ag_2(+)_{589}\text{-}C_4H_4O_6 \longrightarrow$$
$$2AgCl + 2H_2O + (+)_{589}\text{-}[Co(C_2O_4)(en)_2][(+)_{589}\text{-}HC_4H_4O_6] + (-)_{589}\text{-}$$
$$[Co(C_2O_4)(en)_2][(+)_{589}\text{-}HC_4H_4O_6]$$

Separations of diastereoisomers often depend as much on the relative rates of crystallization as on differences in solubilities. The time required for cooling can be critical. Resolution procedures often do not work well if the scale is changed, unless modifications are made.

Procedure

A mixture of 16.0 g (0.050 mole) of $[Co(C_2O_4)(en)_2]Cl{\cdot}H_2O$,* 3.8 g (0.025 mole) of $(+)_{589}$-tartaric acid and 9.1 g (0.025 mole) of disilver $(+)_{589}$-tartrate[11] in 150 mL of hot water is stirred at 60-75° for 15 minutes and then filtered to remove the silver salts, which are washed with about 2 mL of hot water. The combined filtrate and washings are placed immediately in an ice-water bath and stirred with a glass rod to hasten cooling to 10°, which should not require longer than 15 minutes. The sides of the beaker are scratched with a glass rod to induce the formation of crystals. Filtration, using a Büchner funnel, gives filtrate B-1, which is reserved for the recovery of the more-soluble diastereomer below, and dark-red crystals, which are washed successively with 40 mL of 80% aqueous methanol, 40 mL of methanol, and two 40-mL portions of diethyl ether and then dried on the filter by suction. The yield of pure ($\Delta\epsilon_{520}$ = +2.65) $(+)_{589}$-$[Co(C_2O_4)(en)_2]\,[(+)_{589}\text{-}HC_4H_4O_6]{\cdot}H_2O$ is 8.0 g. When less than ideal separations are obtained, the diastereomer may be recrystallized by dissolution in water (9 mL per gram of diastereomer) at 60-70°, followed by cooling the solution to 4° in an ice-water bath. The product is filtered, washed, and dried as before. *Anal.* Calcd. for $[Co(C_2H_8N_2)_2(C_2O_4)]\,[HC_4H_4O_6]{\cdot}H_2O$: C, 27.65; H, 5.30; N, 12.90. Found: C, 27.90; H, 5.16; N, 12.84.

The $(+)_{589}$-isomer is converted to the iodide salt by dissolving the diastereomer in hot water (9 mL/g at 70°) and adding an equimolar amount of NaI (0.346 g per gram of diastereomer). Precipitation begins immediately, and the mixture should be stirred until all the NaI dissolves. The mixture is then cooled in ice to 4° and filtered. The fine orange-red crystals of $(+)_{589}$-$[Co(C_2O_4)(en)_2]I$ are washed with methanol and diethyl ether and dried on the funnel by suction. The yield ($\Delta\epsilon_{520}$ = + 2.64, $[\alpha]_D$ = +720°) from 8.0 g of diastereomer is 7.1 g, 72% of theoretical.

Filtrate B-1, containing the more-soluble diastereomer, is stirred at 60° and 5.0 g of NH_4Br is added, whereupon crystallization of $(-)_{589}$-$[Co(C_2O_4)(en)_2]$-$Br{\cdot}H_2O$ begins. After the NH_4Br dissolves, the mixture is cooled in ice to 30° and then filtered. The red crystals are washed with 80% methanol, pure methanol, and diethyl ether and dried by suction on the filter. The yield ($\Delta\epsilon_{520}$ = − 1.99) is 8.4 g. The bromide salt is then stirred in 165 mL of water at 75° for about 10 minutes, allowed to cool in ice to 15°, and filtered and washed as before. The yield ($\Delta\epsilon_{520}$ = −2.61, $[\alpha]_D$ = 820°) is 6.5 g or 71% of theoretical.

*Prior to the resolution, the $[Co(C_2O_4)(en)_2]Cl{\cdot}H_2O$ should be checked for spectral purity (see Properties, p. 98) and recrystallized if necessary, since impurities interfere with the resolution.

C. RESOLUTION[1] OF POTASSIUM (ETHYLENEDIAMINETETRA-ACETATO)COBALTATE(III) DIHYDRATE, $K[Co(edta)]\cdot 2H_2O$

$$(-)_{589}\text{-}[Co(C_2O_4)(en)_2]^+ + 2[Co(edta)]^- \longrightarrow$$

$$(-)_{589}\text{-}[Co(C_2O_4)(en)_2](-)_{589}\text{-}[Co(edta)] + (+)_{589}\text{-}[Co(edta)]^-$$

Procedure

A dry mixture of 5.0 g (0.030 mole) of silver acetate and 10.9 g (0.030 mole) of $(-)_{589}$-$[Co(C_2O_4)(en)_2]Br\cdot H_2O$ [or 11.8 g (0.030 mole) of $(+)_{589}$-$[Co(C_2O_4)(en)_2]I$] is stirred thoroughly * and then 50 mL of water is added. The mixture is stirred at about 50° for about 15 minutes and filtered, and the residue is washed with 50 mL of hot water. The filtrate (C-1) and washings are added to a solution of 25.4 g (0.0600 mole) of $K[Co(edta)]\cdot 2H_2O$[12] in 100 mL of water at about 50° to precipitate $(-)_{589}$-$[Co(C_2O_4)(en)_2](-)_{546}$-$[Co(edta)]\cdot 3H_2O$. The mixture is cooled in ice with continuous stirring and filtered. The solid diastereomer is washed with 50 mL portions each of ice-cold water, 95% ethanol, absolute ethanol, and diethyl ether and air-dried on the funnel. The yield is about 14 g. The filtrate (C-2) is evaporated to 50 mL to yield a second fraction (1-2 g) and filtrate C-3. The total yield is about 16 g, 80% of theoretical. Precipitation of $K(-)_{546}$-$[Co(edta)]\cdot 2H_2O$ is accomplished by careful addition of about 350 mL of 95% ethanol to filtrate C-3. The yield ($\Delta\epsilon_{585} = +1.53$, $[\alpha]_{546} = -1000°$) is 5.4 g. The diastereomer (16 g) is purified by stirring in 100 mL of water at about 60° and filtering while hot. The yield ($\Delta\epsilon_{544} = -2.11$) is 13.9 g. This is suspended in 50 mL of water, the solution is stirred for 15-20 minutes, and 13.9 g (0.0840 mole) of KI is added to precipitate $(-)_{589}$-$[Co(C_2O_4)(en)_2]I$. The mixture is filtered to give filtrate C-4 and the resolving agent is washed successively with water, 80% methanol, and ether and air-dried. Precipitation of $K(+)_{546}$-$[Co(edta)]\cdot 2H_2O$ occurs upon slow addition of about 350 mL of 95% ethanol to filtrate C-4. The complex is recovered by filtration and washed as before. The yield ($\Delta\epsilon_{585} = +1.50$, $[\alpha]_D = +1000°$) is 8.9 g.

D. POTASSIUM *sym-cis*-(ETHYLENEDIAMINE-*N,N'*-DIACETATO)-DINTROCOBALTATE(III),[2,13] K *sym-cis*-$[Co(edda)(NO_2)_2]$†

$$CoCO_3 + H_2edda + H_2O \longrightarrow [Co(edda)(H_2O)_2] + CO_2$$

*Rapid dissolution of silver acetate is facilitated by mixing with the complex prior to addition of water. Aggregate silver acetate tends to float on the surface, even with vigorous stirring, and dissolves only slowly.

†See reference 8 at end of Section 17 for discussion of isomer descriptions.

$$4[Co(edda)(H_2O)_2] + O_2 + 8KNO_2 \xrightarrow[\text{charcoal}]{\text{activated}} 4K[Co(edda)(NO_2)_2] + 4KOH + 6H_2O$$

Procedure

A mixture of 4.8 g (0.040 mole) of cobalt(II) carbonate and 7.0 g (0.040 mole) of H_2edda in 60 mL of water is stirred at 50° until evolution of gas is no longer observed (about 20 min). Any unreacted $CoCO_3$ is removed by filtration and 7.1 g (0.080 mole) of KNO_2 and 3.0 g of activated charcoal are added. A stream of air is bubbled through the suspension for 12 hours. The air-oxidized mixture is stirred at 60° for a few minutes to dissolve any crystalline product, and the charcoal is removed by filtration. The solution is evaporated using an air stream until crystals have accumulated. The dark red-brown crystals are filtered and washed with 95% ethanol and diethyl ether. Further evaporation of the filtrate yields a second batch of crystalline product. The combined yield is 6 g. Recrystallization from a minimum of hot (about 70°) water by cooling to 20° yields about 5 g of product. *Anal.* Calcd. for $K[Co(C_6H_{10}N_4O_8)]$: C, 19.76; H, 2.77; N, 15.38. Found: C, 19.65; H, 3.12; N, 15.28.

Properties

Only one absorption band appears in the visible region (ϵ_{518} = 151), since the second *d-d* band is covered by a more intense charge-transfer band.

E. RESOLUTION OF POTASSIUM *sym-cis*-(ETHYLENEDIAMINE-*N,N'*-DIACETATO)DINITROCOBALTATE(III), K *sym-cis*-[Co(edda)(NO₂)₂]

$$(-)_{589}\text{-}[Co(C_2O_4)(en)_2]^+ + 2[Co(edda)(NO_2)_2]^- \longrightarrow (-)_{589}\text{-}[Co(C_2O_4)(en)_2](-)\text{-}[Co(edda)(NO_2)_2] + (+)\text{-}[Co(edda)(NO_2)_2]^- *$$

$$(-)_{589}\text{-}[Co(C_2O_4)(en)_2](-)\text{-}[Co(edda)(NO_2)_2] \xrightarrow[\text{ion exchange}]{K^+} K(-)\text{-}sym\text{-}cis\text{-}[Co(edda)(NO_2)_2]$$

*Signs with subscripts give the sign of optical rotation at the wavelength indicated by the subscript, whereas nonsubscripted signs indicate the sign of the dominant circular dichroism peak in the visible region.

Procedure

A mixture of 1.4 g (0.0085 mole) of silver acetate and 3.1 g (0.0085 mole) of $(-)_{589}$-$[Co(C_2O_4)(en)_2]Br \cdot H_2O$ is mixed well with a stirring rod and 35 mL of water is added. The slurry is stirred at 55° for 10 minutes and filtered, and the silver salts are washed with 10 mL of hot water. The filtrate and washings, containing the more soluble acetate salt of the resolving agent, are added to a stirred solution of 6.2 g (0.017 mole) of $K[Co(edda)(NO_2)_2]$ in 50 mL of water. After being stirred a few minutes, the solution is cooled in an ice-water bath to 9°. Formation of red crystals of the diastereomer is induced by scratching the sides of the beaker with a glass rod. These are collected on a filter, washed with 95% ethanol and diethyl ether, and dried by suction. The yield of the less soluble diastereomer ($[\Delta\epsilon]_{508} = -0.635$)* is 3.4 g. The diastereomer is dissolved in 75 mL of water, stirred with 10 g of Dowex 50W-X8 cation-exchange resin in the K^+ form for 15 minutes, and filtered. The solution containing the resolved complex is stirred with an additional 3 g of resin to ensure complete removal of $(-)_{589}$-$[Co(C_2O_4)(en)_2]^+$. The resin is removed by filtration and the filtrate is evaporated to a volume of 20 mL using an air stream. The large dark-red crystals of K (−)-*sym-cis*-$[Co(edda)(NO_2)_2]$ are filtered and washed with 95% ethanol and then diethyl ether. The yield ($\Delta\epsilon_{496} = -2.50$) is 1.6 g.

References

1. W. T. Jordan, B. J. Brennan, L. R. Froebe, and B. E. Douglas, *Inorg. Chem.*, **12**, 1827 (1973).
2. W. T. Jordan and B. E. Douglas, *Inorg. Chem.*, **12**, 403 (1973).
3. C. W. Van Saun and B. E. Douglas, *Inorg. Chem.*, 8, 115 (1969); W. T. Jordan and J. I. Legg., *op. cit.*, **13**, 955 (1974).
4. C. W. Maricondi and B. E. Douglas, *Inorg. Chem.*, **11**, 688 (1972).
5. C. W. Maricondi and C. Maricondi, *Inorg. Chem.*, **12**, 1524 (1973).
6. See, for example, J. Hidaka and Y. Shimura, *Bull. Chem. Soc. Japan*, **40**, 2312 (1967); N. Matsuoka, J. Hidaka, and Y. Shimura, *Inorg. Chem.*, 9, 719 (1970); C. W. Van Saun and B. E. Douglas, *op. ct.*, 8, 115 (1969); W. Byers and B. E. Douglas, *op. cit.*, **11**, 1470 (1972).
7. F. P. Dwyer, I. K. Reid, and F. L. Garvan, *J. Am. Chem. Soc.*, **83**, 1285 (1961).
8. F. P. Dwyer and F. L. Garvan, *Inorg. Synth.*, **6**, 192 (1960).
9. F. P. Dwyer and F. L. Garvan, *Inorg. Synth.*, **6**, 195 (1960).
10. J. A. Broomhead, F. P. Dwyer, and J. W. Hogarth, *Inorg. Synth.*, **6**, 183 (1960).
11. L. J. Halloran, A. L. Gillie, and J. I. Legg, *Inorg. Synth.*, **18**, 103 (1978).
12. F. P. Dwyer, E. C. Gyarfas, and D. P. Mellor, *J. Chem. Phys.*, **59**, 296 (1955).
13. K. Kuroda and K. Watanabe, *Bull. Chem. Soc. Japan.*, **44**, 2550 (1971).

*This $[\Delta\epsilon]_\lambda$ is the *specific, not molar,* circular dichroism. Molar values are not given because the molecular weight of the diastereomer is not known. Since the molar circular dichroism of the diastereomer is assumed to be of little interest, the specific CD should suffice for following a resolution. $[\Delta\epsilon]_\lambda = (\Delta\epsilon)(100)/MW$.

17. ETHYLENEDIAMINE-*N,N'*-DIACETIC ACID COMPLEXES OF COBALT(III)

Submitted by LEON J. HALLORAN,* ARLENE L. GILLIE,* and J. IVAN LEGG*
Checked by PATRICK J. GARNETT,† and DONALD W. WATTS†

The linear quadridentate chelating agent ethylenediamine-*N,N'*-diacetic acid ((ethylenediimino)diacetic acid) (H_2edda = acid, edda = ion)‡ and its *N*-alkylated derivatives have contributed significantly to stereochemical studies involving cobalt(III) complexes. Of particular interest are the investigations correlating alteration in stereochemistry with circular dichroic[1-5] and proton NMR[5-7] spectral properties. In addition, cobalt(III) complexes of edda have played an important role in developing ion-exchange chromatography as a sensitive technique for the separation of isomers possessing subtle structural differences.[7]

Octahedral cobalt(III) complexes with edda and a bidentate ligand (necessarily occupying cis sites) form two geometrical isomers, symmetrical-cis (*sym-cis*), where the two coordinated edda carboxylates are trans and unsymmetrical-cis (*asym-cis*), where the carboxylates are cis.[8] Until recently the *sym-cis*-isomers, obtained in high yield by direct oxidation of cobalt(II) in the presence of edda, were the only readily available isomers. The published syntheses for *asym-cis*-$[Co(edda)(diamine)]^+$ complexes have given very low yields[9] or have been nonreproducible,[10] limiting study of these potentially interesting compounds. However, in contrast to the behavior with diamines, dicarboxylato ligands such as carbonato or oxalato give large amounts of the *asym-cis*-isomers.[6,11] Because of the ease of replacement of the coordinated carbonate, $Na[Co(edda)(CO_3)]$ can serve as the general starting material for the synthesis of a number of asym-cis as well as sym-cis Co(III) edda complexes.[7,12]

Presented here are the synthesis and resolution of *asym-cis*- and *sym-cis*-$[Co(edda)(en)]^+$, as well as the synthesis of $Na[Co(edda)(CO_3)]$ (primarily the *asym-cis*-isomer) employed for the synthesis of *asym-cis*-[Co(edda)(en)]Cl. For the preparation of asym-cis complexes there is no advantage in obtaining pure *asym-cis*-$[Co(edda)(CO_3)]^-$, since some isomerization occurs during the displacement of the carbonate, necessitating subsequent separation of isomers. However, the two isomers of $[Co(edda)(CO_3)]^-$ can be obtained readily by fractional crystallization.[12] These synthetic, separation, and resolution procedures are generally useful for charged complexes, such as those of amino acids.

*Department of Chemistry, Washington State University, Pullman, WA 99164.
†School of Chemistry, The University of Western Australia, Nedlands, Western Australia 6009.

‡Lower case letters are used for edda as recommended by the I.U.P.A.C. nomenclature rules, even though this is not current practice in the literature.

A. SODIUM[CARBONATO[ETHYLENEDIAMINE-N,N′-DIACETATO-(2−)]COBALTATE(III)]

$$Na_3[Co(CO_3)_3] + H_2edda + 2HCl \longrightarrow Na[Co(edda)(CO_3)] + 2CO_2 + 2H_2O + 2NaCl$$

edda = ethylenediamine-*N,N′*-diacetate ion

The synthesis of $[Co(edda)(CO_3)]^-$ was first reported by Mori et al.[14] but was found to be nonreproducible by Van Saun and Douglas.[15] Recent reports by Garnett et al.[11,12] show that the isomeric composition is primarily asym-cis. All the reported syntheses are based on the reaction of $[Co(CO_3)_3]^{3-}$ with the free acid H_2edda. The modification employed in this synthesis involves the addition of sufficient HCl to prevent the formation of $NaHCO_3$, which may contaminate the product recovered from the Garnett and Watts synthesis.[12]

Procedure

To a slurry of 25.9 g (0.0715 mole) of freshly prepared $Na_3[(Co(CO_3)_3]\cdot 3H_2O$[13] in 30 mL of water is added 12.65 g (0.0715 mole) of ethylenediamine-*N,N′*-diacetic acid,* and the solution is heated to 50° with stirring for 10 minutes. Ten milliliters of concentrated HCl diluted to 30 mL is added. The solution is filtered and cooled in an ice bath. The product is precipitated by the slow (20 min) addition of 100 mL absolute ethanol to the stirred solution. The dark-purple complex is filtered, washed with 50% aqueous ethanol and ethanol and air-dried. The total yield is 16.4 g (67%). The crude product is used without further purification in the synthesis below. A single recrystallization from ethanol-water gives a pure product. *Anal.* Calcd. for $Na[CoC_6H_{10}N_2O_7]\cdot 1.5H_2O$: C, 24.50; H, 3.82; N, 8.16. Found: C, 24.64; H, 3.79; N, 8.17.

Properties

The compound is stable in the solid state indefinitely. The optical isomers of both geometric isomers have been resolved but racemize in solution (some loss in optical activity is observed within 5 min.).[1,12,15]

*LaMont Laboratories, Inc., P. O. Box 79, Tyngsboro, MA 01879.

B. *asym-cis*-[(ETHYLENEDIAMINE) [ETHYLENEDIAMINE-*N,N'*-DIACETATO(2−)]COBALT(III)] CHLORIDE

$$\text{Na[Co(edda)(CO}_3\text{)]} + 2\text{H}_2\text{O} + 2\text{HCl} \longrightarrow$$
$$\textit{asym-cis-}\text{[Co(edda)(H}_2\text{O)}_2\text{]Cl} + \text{CO}_2 + \text{NaCl} + \text{H}_2\text{O}$$

$$\textit{asym-cis-}\text{[Co(edda)(H}_2\text{O)}_2\text{]Cl} + \text{en} \longrightarrow$$
$$\textit{asym-cis-}\text{[Co(edda)(en)]Cl} + 2\text{H}_2\text{O}$$

The experimental procedure is based on the removal of coordinated carbonate by the addition of hydrochloric acid to form the diaqua complex, with subsequent coordination of the diamine. Excess acid is used to inhibit isomerization of the *asym-cis*-diaqua complex to the sym-cis form. Although the *asym-cis*-isomer makes up the majority of the starting material, some isomerization occurs, resulting in the formation of a significant amount of the *sym-cis*-isomer. Isomer separation is achieved by fractional crystallization or by cation-exchange chromatography. If there is any question about the purity of the isomer obtained by fractional crystallization, the chromatographic method should be used.

Procedure

To 7.9 g (0.023 mole) of Na[Co(edda)(CO_3)]·1.5H_2O in 40 mL of water is added 6 mL of concentrated hydrochloric acid (0.06 mole). When the evolution of gas has ceased, 3 g (0.05 mole) of ethylenediamine is added to the red-violet solution. After 1 hour the *asym-cis*-product is removed by filtration (on standing longer the *sym-cis* may begin to precipitate), washed with 50% aqueous ethanol, ethanol, and acetone, and air-dried. The yield is 1.8 g.

Attempts to separate more of the *asym-cis*-isomer by further volume reduction result in the formation of a thick, fibrous mat of *sym-cis*-[Co(edda)(en)]Cl crystals. If desired, the rest of the *asym-cis* complex can be separated from the *sym-cis* isomer by using a very short (4.5-cm diameter, 3.0-cm length) column of Bio-Rad AG 50W-X8,* 50-100 mesh, Na^+ form (obtained by passing 250 mL of 1.5 *M* NaCl through the resin in the acid form followed by 100 mL of water). The reaction solution is chromatographed in portions. The size of the portion is determined by the height of the band, which should not exceed ½ cm. The band is eluted with 0.5 *M* NaCl solution at 2 mL/min. The material is removed in two fractions, a faster-moving, red band of the *sym-cis*-isomer and a slower-moving,

*The resin is available from Bio-Rad Laboratories, Richmond, CA 94804.

orange band of the *asym-cis*-isomer. The second band is collected, combined with the other second-band fractions (obtained from elution of the remaining portions), and evaporated under an airstream until a wet solid mass of NaCl is obtained. Absolute ethanol (two to three times the volume of eluate) is added to precipitate additional salt, the mixture is filtered, and the salt is washed with 50% aqueous ethanol until almost white. The process is repeated on the filtrate and washings until treatment with ethanol begins to cause the complex to precipitate as a red powder. At this point the solution is evaporated to a small volume to remove ethanol and filtered to remove any precipitated NaCl. The final desalting is accomplished by gel permeation chromatography. The filtrate is layered on a column (1.5 × 40 cm) of Sephadex G-10* and eluted with water (0.5 mL/min) saturated with $CHCl_3$ (added to prevent bacterial growth). The solid [Co(edda)-(en)]Cl·$3H_2O$ is isolated by evaporating the eluant until crystals begin to form, warming the solution to redissolve the solid, adding a small amount of ethanol (about 10% of the total volume), and cooling the mixture. The combined yield of *asym-cis*-[Co(edda)(en)]Cl·$3H_2O$ is 3.1 g (35%). *Anal.* Calcd. for $[CoC_8H_{18}N_4O_4]Cl\cdot 3H_2O$: C, 25.10; H, 6.32; N, 14.64. Found: C, 25.10; H, 6.17; N, 14.90.

Properties

The dark-red crystals of *asym-cis*-[Co(edda)(en)]Cl dissolve in water at room temperature to give a solution stable to isomerization. The visible absorption spectrum in water shows λ_{max} (ϵ) = 493 (170), 359 nm (169). The proton NMR spectrum of the complex has been assigned.[6] The isomer is characterized by its acetate methylene proton resonances, which occur at 4.22, 3.92, 3.71, and 3.40 ppm from the internal reference sodium 3-(trimethylsilyl)-1-propanesulfonate.

C. RESOLUTION OF *asym-cis*-[(ETHYLENEDIAMINE) [ETHYLENE-DIAMINE-*N,N'*-DIACETATO(2−)]-COBALT(III)] CHLORIDE

asym-cis-[Co(edda)(en)]Cl + Ag(brcamsul) ⟶

$(-)_{485}$- and $(+)_{485}$-*asym-cis*-[Co(edda)(en)](brcamsul) + AgCl

$(-)_{485}$- or $(+)_{485}$-*asym-cis*-[Co(edda)(en)](brcamsul) + ®-Cl ⟶

$(-)_{485}$- or $(+)_{485}$-*asym-cis*-[Co(edda)(en)]Cl + ®-(brcamsul)

(edda = ethylenediamine-*N,N'*-diacetate, en = ethylenediamine, brcamsul = (+)-α-bromocamphor-π-sulfonate ([1*R*-(*endo,anti*)]-3-bromo-1,7-dimethyl-2-oxobi-

*The resin is available from Bio-Rad Laboratories, Richmond, CA 94804.

cyclo[2.2.1]heptane-7-methanesulfonate), ®-Cl = Bio-Rad AG 1-X8 anion-exchange resin in the Cl^- form, see footnote p. 106.)

Although the diastereomers do not differ significantly in solubility, the distinct crystal habits greatly facilitate separation by fractional crystallization.[7]

Procedure

The silver (+)-α-bromocamphor-π-sulfonate monohydrate[16] used for the resolution is prepared by dissolving 4.1 g (0.0125 mole) of NH_4(brcamsul) (Aldrich) and 2.1 g (0.0125 mole) of $AgNO_3$ in 40 mL water at 70°. The hot solution is filtered and evaporated in an airstream to 30 mL, reheated to dissolve any precipitated product, and cooled in a refrigerator overnight. The small white needles of Ag(brcamsul)·H_2O are filtered, washed with 1:1 ethanol-diethyl ether and diethyl ether, and air-dried. The yield ranges from 2.5 to 3 g.

The complex *asym-cis*-[Co(edda)(en)]Cl·$3H_2O$, 1.79 g (0.0047 mole) is dissolved in 40 mL of water by heating to 60° and 2.04 g (0.0047 mole) Ag(brcamsul)·H_2O is added. Additional small increments are added until no further AgCl precipitate is obtained. A large excess of the resolving agent should be avoided because of decomposition in subsequent operations to give black silver deposits on the glassware. The solution is maintained at 60° for ½ hour to coagulate the silver chloride and is then filtered. The silver chloride is washed with several small portions of water until the washings are colorless. The filtrate and washings are fractionally crystallized by evaporation under an airstream followed by refrigeration. The diastereomers are easily distinguished, the (−)-diastereomer forming massive, rectangular crystals while the (+)-diastereomer forms a thick, cottonlike mass of fine needles. The diastereomers precipitate in random order, but each fraction consists of one nearly pure diastereomer. If mixed crystals are obtained, the concentration of the crystallizing solution is too high. The crystals are then redissolved in a larger amount of water and recrystallized. The combined fractions of each diastereomer are recrystallized by dissolving the solid in a minimum amount of hot water (60-70°), cooling the solution in a refrigerator, and filtering. The crystals are washed with 50% ethanol-water and absolute ethanol and air-dried. Two recrystallizations are sufficient to give diastereomers of contstant rotation. $\Delta\epsilon$ values, calculated assuming anhydrous salts, and yields are: $(-)_{485}$-*asym-cis*-[Co(edda)(en)](brcamsul), $\Delta\epsilon_{485} = -1.85$ (1.12 g, 77%); $(+)_{485}$-*asym-cis*-[Co(edda)(en)](brcamsul), $\Delta\epsilon_{485} = +2.05$ (0.75 g, 52%).

Chloride salts of the optical isomers are prepared by removing the resolving agent with anion-exchange resin. A 0.47-g sample of $(-)_{485}$-*asym-cis*-[Co(edda)-(en)](brcamsul) is dissolved in 10 mL of water and 4.0 g of washed Bio-Rad AG 1-X8 anion-exchange resin, 100-200 mesh, Cl^- form, is added. The slurry is allowed to sit with periodic stirring for several hours. The resin is removed by filtration and washed until the washings are colorless. Another 4.0 g batch of resin is added to the combined filtrate and washings, and the above procedure is repeated. The filtrate and washings are concentrated under an airstream to 10 mL. The solution is heated to dissolve the solid that forms, and while still hot, 5 mL of absolute ethanol is added. The mixture is cooled in a refrigerator and filtered, and the product is washed with 50% aqueous ethanol and absolute ethanol to yield 0.22 g (74%) of thin, red platelets. The other optical isomer is recovered in the same way—0.32 g of diastereomer gives 0.18 g (89%) of the chloride salt. *Anal.* Calcd. for $[CoC_8H_{18}N_4O_4]Cl \cdot 3H_2O$: C, 25.11; H, 6.32; N, 14.64. Found for $(-)_{485}$-isomer: C, 25.08; H, 6.04; N, 14.75. Found for $(+)_{485}$-isomer: C, 25.06, H, 5.90; N, 14.69.

Properties

Aqueous solutions of the enantiomers of *asym-cis*-[Co(edda)(en)]Cl are stable to racemization at room temperature. The circular dichroism spectrum of the (−) isomer in water shows λ_{max} ($\Delta\epsilon$): 485 nm(−2.24), 357 nm (+0.89). The absolute configuration of the closely related (*R*)-1,2-diaminopropane [(−)-pn] isomer, *asym-cis*-[Co(edda)[(−)-pn]]Cl, has been determined by x-ray crystallography.[17] The results of this study confirm the absolute configuration assigned to the two *asym-cis*-[Co(edda)(en)]Cl enantiomers on the basis of their circular-dichroism spectra.[7]

D. *sym-cis*-[(ETHYLENEDIAMINE)[ETHYLENEDIAMINE-*N,N′*-DIACETATO(2−)]COBALT(III)] NITRATE

$$4CoCO_3 + 4H_2edda + 4en + 4HNO_3 + O_2 \xrightarrow{C}$$

$$4\textit{sym-cis-}[Co(edda)(en)]NO_3 + 4CO_2 + 6H_2O$$

edda = ethylenediamine-*N,N′*-diacetate ion, en = ethylenediamine

Since the *sym-cis*-isomer is favored thermodynamically, it can be isolated in high yields by direct oxidation of Co(II) in the presence of the ligands and decolorizing carbon as a catalyst.[9]

Procedure

A suspension of 13.1 g (0.11 mole) of cobalt(II) carbonate and 17.6 g (0.10 mole) of ethylenediamine-*N,N′*-diacetic acid in 250 mL of water is heated at 60°

with occasional stirring until the carbon dioxide evolution ceases (about 20 min). The pink solution is filtered through a medium fritted glass filter, and the unreacted $CoCO_3$ is washed with 150 mL of water. To the combined filtrate and washings are added successively 50 mL of 2 *M* nitric acid, 10 g of decolorizing carbon (Norit-A, alkaline form, from Fisher), and 6.1 g (0.10 mole) of 98% ethylenediamine in 40 mL of water. Air is bubbled through the mixture for 8 hours and the charcoal is removed by filtration. Evaporation of the solution on a steam bath to about 50 mL yields red-violet crystals, which are filtered, washed with three 10-mL portions of water, 50% ethanol, ethanol, and acetone, and air-dried; the yield of *sym-cis*-[Co(edda)(en)]$NO_3 \cdot H_2O$ is 24 g (65%). *Anal.* Calcd. for [$CoC_8H_{18}N_4O_4$]$NO_3 \cdot H_2O$: C, 25.74; H, 5.40; N, 18.77. Found: C, 25.86; H, 5.63; N, 18.60. More product can be obtained by evaporation of the combined filtrate and washings.

Properties

The dark red-violet crystals of *sym-cis*-[Co(edda)(en)]NO_3 dissolve in water at room temperature to give a solution stable to isomerization. The visible absorption spectrum in water shows λ_{max} (ϵ) = 529 (87.3), 448 (shoulder), 362 nm (113). The proton NMR spectrum of the complex has been assigned.[9] The isomer is characterized by its acetate methylene proton resonances, which occur at 4.37, 4.07, 3.49, and 3.20 ppm from the internal reference sodium 3-(trimethylsilyl)-1-propanesulfonate.

E. RESOLUTION OF *sym-cis*-[(ETHYLENEDIAMINE)[ETHYLENEDIAMINE-*N,N'*-DIACETATO(2−)]COBALT(III)] NITRATE

$$sym\text{-}cis\text{-}[Co(edda)(en)]NO_3 + \text{®-}Cl \longrightarrow sym\text{-}cis\text{-}[Co(edda)(en)]Cl + \text{®-}NO_3$$

$$2sym\text{-}cis\text{-}[Co(edda)(en)]Cl + Ag_2(tartrate) + H_2(tartrate) \longrightarrow 2(-)_{529}\text{- and } (+)_{529}\text{-}sym\text{-}cis\text{-}[Co(edda)(en)]H(tartrate) + 2AgCl$$

$$(-)_{529}\text{- or } (+)_{529}\text{-}sym\text{-}cis\text{-}[Co(edda)(en)]H(tartrate) + HNO_3 \longrightarrow (-)_{529}\text{- or } (+)_{529}\text{-}sym\text{-}cis\text{-}[Co(edda)(en)]NO_3 + H_2tartrate$$

edda = ethylenediamine-*N,N'*-diacetate ion, en = ethylenediamine, ®–Cl= Bio-Rad AG 1-X8 anion-exchange resin in the Cl^- form, see footnote, p. 106.

Since there are problems with the isolation of the silver hydrogen tartrate employed in the resolution,[18] a mixture of the correct stoichiometry is obtained by mixing disilver tartrate with tartaric acid.

Procedure

Disilver tartrate is prepared in the following manner. (During the synthesis the reaction mixture is shielded from light.) A solution of 3.8 g (0.025 mole) of (+)-tartaric acid in 15 mL of water, to which has been added 2 g (0.050 mole) of 98% sodium hydroxide in 10 mL of water, is added dropwise to a stirred solution of 8.5 g (0.050 mole) of silver nitrate in 25 mL of water. After about 30 minutes of stirring, the precipitate is allowed to settle, filtered, and washed with five 30-mL portions of water and then acetone. The air-dried disilver tartrate (7.6 g) is stored in a brown bottle.

The disilver tartrate is used to resolve the complex as follows. A solution of 7.464 g (0.02 mole) of *sym-cis*-[Co(edda)(en)]$NO_3 \cdot H_2O$ in 160 mL of water is passed through a column (diameter 2.8 cm) that contains 200 mL (wet volume) of Bio-Rad 1-X8 strong-base anion-exchange resin (50-100 mesh) in the chloride form at a rate between 1 and 2 mL/min. The eluted complex is collected quantitatively in a stirred suspension of 3.638 g (0.01 mole) of disilver tartrate and 1.501 g (0.01 mole) of tartaric acid in 100 mL of water, which is kept shielded from light. The coagulated silver chloride is then filtered but not washed, and the filtrate is evaporated to about 50 mL with air alone. To the stirred solution at 40° is then added 15 mL of ethanol. Within a few minutes crystallization takes place. After 15 minutes of stirring, 10 mL of water is added and the mixture is stirred an additional 15 minutes. The pure $(-)_{532}$-diastereomer is then filtered at 40°, washed with two 5-mL protions of 60% aqueous ethanol (these washings are added to the filtrate containing the $(+)_{532}$-diastereomer), ethanol and diethyl ether, and air-dried. Yield of $(-)_{532}$-*sym-cis*-[Co(edda)(en)]H(tartrate)$\cdot 0.5H_2O$ is 1.6 g (23%) ($\Delta\epsilon_{532}$ = −4.45). Recrystallization from 50% ethanol does not change the rotation. *Anal.* Calcd. for $[CoC_8H_{18}O_4N_4]$-$(C_4H_5O_6)\cdot 0.5H_2O$: C, 31.94; H, 5.36; N, 12.42. Found: 31.78; H, 5.45; N, 12.54.

To isolate the (+)-diastereomer, 5 mL of water is added to the combined filtrates obtained from the isolation of the (−)-diastereomer to dissolve any precipitate. The solution is then cooled in an ice bath, and the almost pure (+)-diastereomer is filtered and washed as described for the (−)-isomer to yield 3.2 g. The product is then dissolved in 35 mL of water at 40°, and 30 mL of ethanol is added. The solution is cooled in an ice bath, and the product is filtered and washed as before. Yield of $(+)_{532}$-*sym-cis*-[Co(edda)(en)]H(tartrate)$\cdot 0.5H_2O$ is 1.7 g (24%) ($\Delta\epsilon_{532}$ = +4.46).

The (+)-diastereomer (1.7 g) is dissolved in 6 mL of water, and 1 mL of concentrated HNO_3 is added. The nitrate salt is precipitated by the slow addition of 15 mL of ethanol to the stirred solution The product is filtered, washed with 70% aqueous ethanol, ethanol, and diethyl ether, and air-dried. The yield of $(+)_{532}$-*sym-cis*-[Co(edda)(en)] $NO_3 \cdot H_2O$ is 1 g(70%). *Anal.* Calcd. for $[CoC_8H_{18}N_4O_4]$-$NO_3 \cdot H_2O$: C, 25.74; H, 5.40; N, 18.77. Found: C, 25.70; H, 5.50; N, 18.72. The (–)-diastereoisomer (1.6 g) is converted similarly to the nitrate using 8 mL of water, 1 mL of concentrated HNO_3, and 15 mL of ethanol. The yield of $(-)_{532}$-*sym-cis*-[Co(edda)(en)] $NO_3 \cdot H_2O$ is 0.6 g (45%).

Properties

Aqueous solutions of the enantiomers of *sym-cis*-[Co(edda)(en)] NO_3 are stable to racemization at room temperature. The circular dichroism spectrum of the (+)-isomer shows λ_{max} ($\Delta\epsilon$) = 532 (+4.46) and 446 nm (–1.75), and two very broad and weak bands below 400 nm.

References

1. W. T. Jordan and J. I. Legg, *Inorg. Chem.*, **13**, 955 (1974).
2. W. T. Jordan and B. E. Douglas, *Inorg. Chem.*, **12**, 403 (1973).
3. C. W. Maricondi and C. Maricondi, *Inorg. Chem.*, **12**, 1524 (1973).
4. C. W. Maricondi and B. E. Douglas, *Inorg. Chem.*, **11**, 668 (1972).
5. G. R. Brubaker, D. P. Schaefer, J. H. Worrell, and J. I. Legg, *Coord. Chem. Rev.*, **7**, 161 (1971).
6. P. F. Coleman, J. I. Legg, and J. Steele, *Inorg. Chem.*, **9**, 937 (1970).
7. L. J. Halloran and J. I. Legg, *Inorg. Chem.*, **13**, 2193 (1974).
8. The original nomenclature proposed for edda complexes was based on the relative positions of the edda carboxylates.[9] Thus, sym-cis and asym-cis correspond to trans and cis in the original nomenclature. The nomenclature employed here is based on the relative positions of the two remaining sites not occupied by edda and the symmetry of the chelated edda.[2] The α,β designation, corresponding to sym-cis, uns-cis respectively, has been employed also for edda complexes.[12]
9. J. I. Legg and D. W. Cooke, *Inorg. Chem.*, **4**, 1576 (1965).
10. K. Kuroda, *Bull. Chem. Soc. Japan*, **45**, 2176 (1972).
11. P. J. Garnett, D. W. Watts, and J. I. Legg, *Inorg. Chem.*, **8**, 2534 (1969).
12. P. J. Garnett and D. W. Watts, *Inorg. Chim. Acta*, **8**, 293 (1974).
13. H. F. Bauer and W. C. Drinkard, *Inorg. Synth.*, **8**, 202 (1966).
14. M. Mori, M. Shibata, E. Kyuno, and F. Maruyama, *Bull. Chem. Soc. Jap.*, **35**, 75 (1962).
15. C. Van Saun and B. E. Douglas, *Inorg. Chem.*, **8**, 115 (1969).
16. F. Hein and K. Vogt, *Chem. Ber.*, **98**, 1691 (1965).
17. L. J. Halloran, R. E. Caputo, R. D. Willett, and J. I. Legg, *Inorg. Chem.*, **14**, 1762 (1975).
18. J. I. Legg, D. W. Cooke, and B. E. Douglas, *Inorg. Chem.*, **6**, 700 (1967).

18. TRANSITION METAL COMPLEXES OF BIS(TRIMETHYL-SILYL)AMINE (1,1,1,3,3,3-Hexamethyldisilazane)

Submitted by DONALD C. BRADLEY* and RICHARD G. COPPERTHWAITE†
Checked by M. W. EXTINE,‡ W. W. REICHERT,‡ and MALCOLM H. CHISHOLM‡

The compounds $M\{N[Si(CH_3)_3]_2\}_3$ may be considered as a special class of transition metal dialkylamides. The bulky trimethylsilyl groups eliminate intermolecular association and lead to unusually low coordination around the metal atom. These coordinatively unsaturated molecules are highly reactive and may have catalytic applications.

Bürger and Wannagat[1,2] reported the synthesis, on a milligram scale, of the tervalent metal derivatives $Fe\{N[Si(CH_3)_3]_2\}_3$ and $Cr\{N[Si(CH_3)_3]_2\}_3$. We include full details of the preparation, with higher yields of these compounds in sufficient quantities to facilitate full characterization by physical methods. The preparation of the corresponding scandium, titanium, and vanadium analogues are described also.

Although the following methods involve the synthesis of a series of essentially similar complexes, each preparation contains significantly different details in the way of starting materials, solvents, and physical conditions. Thus it was shown[3] that the formation of the trimethylamine adducts $MCl_3{\cdot}2N(CH_3)_3$ was an essential prerequisite for the synthesis of $Ti\{N[Si(CH_3)_3]_2\}_3$ and $V\{N[Si(CH_3)_3]_2\}_3$. Because of this, the preparation times for Ti(III) and V(III) derivatives (4-5 days) are longer than for the Sc(III), Cr(III) and Fe(III) derivatives, (1-2 days).

Procedure

General Techniques. These compounds are highly reactive to moisture and oxygen. Consequently, syntheses and spectral measurements are performed in an all-glass preparative line, operated at a reduced pressure of 0.005 torr and maintained with a single-stage rotary pump connected to a liquid nitrogen trap (Figure 1). High-purity nitrogen is passed through a Pyrex column, 60 cm long by 4 cm in diameter, containing a mixture of reduced oxides of manganese, to remove oxygen. The preparation and regeneration of this type of oxygen remover have been described elsewhere.[4] Final drying is achieved by passage

*Department of Chemistry, Queen Mary College, Mile End Road, London El 4NS, U.K.
†National Chemical Research Laboratory, P.O.B. 395, Pretoria 0001, S. Africa.
‡Department of Chemistry, Princeton University, Princeton, N. J. 08540.

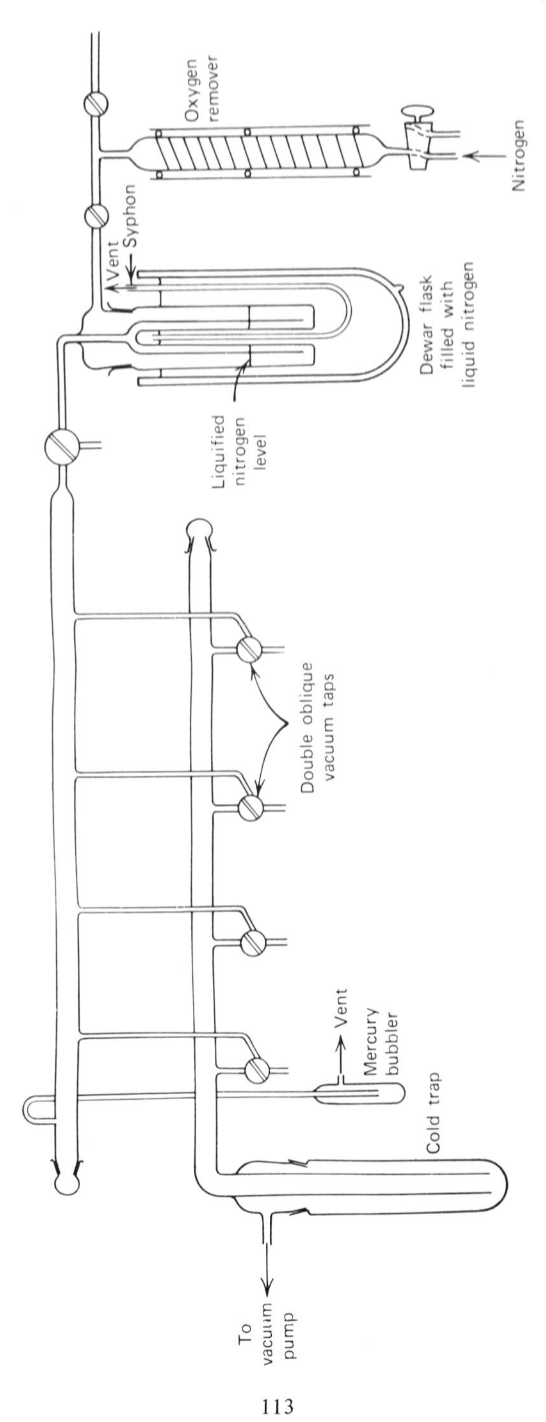
Oxygen
remover
Nitrogen
Vent
Syphon
Dewar flask
filled with
liquid nitrogen
Liquified
nitrogen
level
Double oblique
vacuum taps
Vent
Mercury
bubbler
Cold trap
To
vacuum
pump

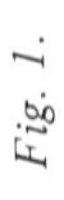
Fig. 1.

through a double-walled condenser of approximately 1-liter capacity immersed in a large (30 × 10 cm diameter) Dewar flask of liquid nitrogen. Sufficient gas pressure is generated in the system to ensure that a 15-cm column of liquid nitrogen remains in equilibrium within the double-walled condenser. By means of two-way taps built into the line, the vacuum or a positive pressure of dry oxygen-free nitrogen is available for use. PVC tubing (0.6 cm inner diameter) reinforced with Nylon is used to connect the preparative line to the glass apparatus. Prior to use, the Pyrex glassware, fitted with standard ground-glass joints, is rinsed in acetone and dried at 110° in an oven [it is in fact preferable to use anhydrous ethanol (distilled) containing a few percent of benzene for drying purposes, since the ternary water-alcohol-benzene azeotrope is the most volatile component, and thus removes water more readily, whereas acetone is much more volatile than water with which it does not form an azeotrope]. On connection to the preparative line the apparatus is pumped under vacuum while the outside is flamed with a large Bunsen flame. The object is to heat all parts of the apparatus to a point somewhat below the annealing temperature so that the flame assumes a distinct yellow color owing to the sodium in the glass. By this means any adsorbed oxygen on the insides of flasks and ampules is pumped away. (Care must be taken to avoid local heating of any particular region that might lead to softening and consequent distortion as a result of the pressure difference inside and outside the apparatus). With this preparative line the usual techniques of filtration, distillation, sublimation, and recrystallization can be performed using fairly standard Pyrex equipment. For filtrations and the storage of air-sensitive products, apparatus similar to that described by R. J. H. Clark[5] is used. All chemicals are transferred against a brisk countercurrent of dry nitrogen—*solids* by the use of transfer tubes previously loaded in a nitrogen-purged glove box and *liquids* with syringes.

The preparation of some reagents that are used as starting materials for the procedures given below can be discussed at this point. *1,1,1,3,3,3-Hexamethyldisilazane,* $[(CH_3)_3Si]_2NH$, is prepared[6] in 500-mL amounts and stored over activated molecular sieves. *Lithium bis(trimethylsilyl)amide,* $LiN[Si(CH_3)_3]_2$, is prepared by a method similar to that previously given[7], under a dry nitrogen atmosphere. Excess of the disilazane (about 5%) is added slowly, with ice cooling, to a 10% w/v solution of *n*-butyllithium (1.56 *M*) in a mixed hydrocarbon solvent (purchased from Metallgesellschaft A.G., Frankfurt). The lithium derivative is used directly in solution, or, by stripping the solvent, the solid can be stored under dry nitrogen and added by means of transfer tubes when required.

A. TRIS[BIS(TRIMETHYLSILYL)AMIDO]SCANDIUM(III) (tris(1,1,1,3,3,3-hexamethyldisilazanato)scandium(III))

$$ScCl_3 + 3LiN[Si(CH_3)_3]_2 \longrightarrow Sc\{N[Si(CH_3)_3]_2\}_3 + 3LiCl$$

Procedure

Freshly distilled bis(trimethylsilyl)amine (20 mL, 0.10 mole) contained in a 50-mL dropping funnel, is added dropwise to 60 mL, (0.093 mole) of the standard *n*-butyllithium solution (see General Technique section), contained in a 250-mL, round-bottomed flask immersed in an ice-water bath, as shown in Figure 2. The addition should take about 1 hour, after which time white crystals

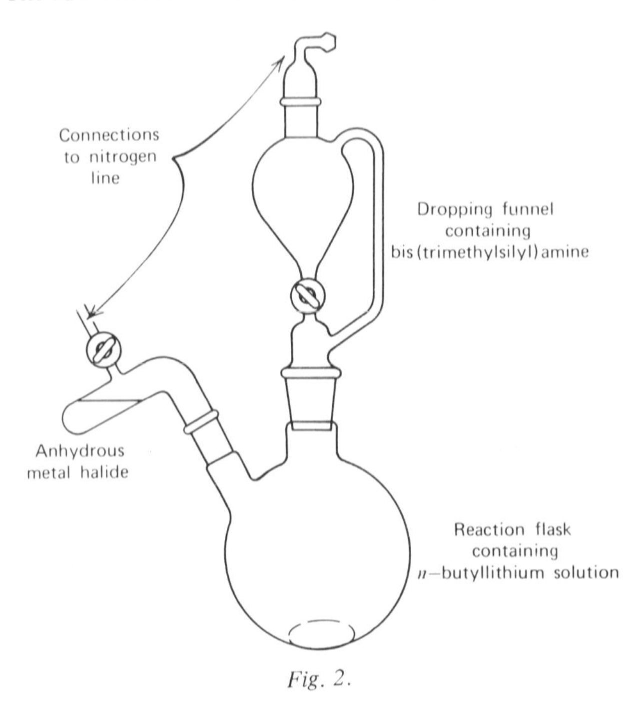

Fig. 2.

of the lithium salt separate as a slurry. Dry tetrahydrofuran, 100 mL, previously distilled over KOH pellets and stored over calcium hydride, is added by syringe against a countercurrent of dry nitrogen. All the solid dissolves with warming to give a pale-yellow solution. Anhydrous scandium trichloride (4.60 g, 0.030 mole), prepared by treating the hexahydrate with refluxing sulfinyl chloride ($SOCl_2$), is added slowly with ice cooling over about 30 minutes by turning the transfer tube in the socket and tapping gently. The mixture is then allowed to stir at room temperature for 24 hours. All the solvent is stripped and vacuum pumping on the dry residue is continued for 6 hours. Dry pentane, 100 mL, freshly distilled and stored over calcium hydride, is added by means of the syringe and the mixture is warmed gently with an infrared lamp to give a white precipitate and a pale solution. The white residue is removed by filtration using equipment of the type described in reference 5. Concentration of the pentane solution yields thin colorless needles, which are filtered, transferred in a glove box to a glass manifold, and sealed under vacuum in Pyrex ampules. The yield of recrystallized tris[bis(trimethylsilyl)amido]scandium(III) is about 8 g (50%). *Anal.* Calcd. for $Sc\{N[Si(CH_3)_3]_2\}_3$: C, 41.09; H, 10.35; N, 7.99. Found: C, 40.50; H, 10.11; N, 7.69.

Properties

Tris[bis(trimethylsilyl)amido]scandium(III) crystallizes from pentane as white, fragile needles, very sensitive to moisture, mp 172-174° (sealed tube). The proton NMR spectrum of a 10% solution in benzene, (TMS as internal standard) gives a single sharp resonance at τ 9.66 and the mass spectrum shows a parent molecular ion corresponding to the three-coordinate monomer. Some infrared and electronic absorption spectra are given in Table II. The single-crystal X-ray

TABLE II Physical Properties of $M\{N[Si(CH_3)_3]_2\}_3$ Derivatives

		Infrared Spectra (cm^{-1})[b]	
M	Electronic Spectra (cm^{-1})[a]	M–N stretching	N–Si stretching
Sc	31,200; 40,800	382,420	950
Ti	4800; 17,400; 28,600	380	899
V	12,000; 15,900; 24,700; 28,100	418	902
Cr	11,800; 14,800; 25,300; 31,400	376	902
Fe	16,100; 20,000; 25,300, 29,700	376	902

[a]Data from reference 10, which also includes magnetic susceptibilities, electron paramagnetic resonance data, and crystal field calculations.
[b]Data are characteristic bands taken from reference 9, which also includes mass spectral data.

structural analysis shows that this compound has a pyramidal ScN_3 configuration in the crystalline state, thus differing from the Ti, V, Cr, and Fe derivatives.[13]

B. TRIS[BIS(TRIMETHYLSILYL)AMIDO]TITANIUM(III)(Tris(1,1,1,3,3,3-hexamethyldisilazanato)titanium(III))

$$TiCl_3 + 2N(CH_3)_3 \longrightarrow TiCl_3 \cdot 2N(CH_3)_3$$

$$TiCl_3 \cdot 2N(CH_3)_3 + 3LiN[Si(CH_3)_3]_2 \longrightarrow$$
$$Ti\{N[Si(CH_3)_3]_2\}_3 + 3LiCl + 2N(CH_3)_3$$

Procedure

Anhydrous trimethylamine (125 mL) is distilled from P_4O_{10} onto 5.0 g (0.032 mole) of titanium trichloride (Alfa Inorganics) in a 250-mL, two-necked, round-bottomed flask equipped with a "cold-finger condenser" charged with Dry Ice-acetone. With this arrangement the trimethylamine can be maintained at reflux without any external cooling of the reaction flask. Stirring is continued for 48 hours (this includes two overnight sessions when the whole flask is immersed in a large Dewar flask of Dry Ice-acetone and stored under nitrogen) after which large quantities of the blue amine adduct are formed (soluble in the excess trimethylamine). At this stage 16.03 g (0.06 mole) of lithium bis(trimethylsilyl)-amide is added, through a transfer tube over a period of 1 hour, and the mixture is stirred at reflux for a further 72 hours. The solvent is then stripped off and the dark-blue solid residue is subjected to vacuum pumping for 8 hours. Dry pentane (100 mL, previously distilled from calcium hydride and degassed three times under nitrogen) is added, and a royal blue solution forms, together with a white precipitate. The solid is filtered in the usual way and concentration of the solution yields a mass of bright-blue needle-shaped crystals. These are filtered, washed with pentane, dried, and transferred under vacuum into Pyrex ampules, which are then sealed. The yield of tris[bis(trimethylsilyl)amido]titanium(III) is about 10 g (60%). *Anal.* Calcd. for $Ti\{N[Si(CH_3)_3]_2\}_3$: Ti, 9.05; C, 40.86; H, 10.29; N, 7.94. Found: Ti, 8.65; C, 39.91; H, 9.92; N, 7.64.

Properties

Tris[bis(trimethylsilyl)amido]titanium(III) crystallizes from pentane as bright-blue needles, extremely sensitive to oxygen and moisture. Both the solid and hydrocarbon solutions of this compound give strong electron paramagnetic resonance signals at room temperature. The g-values obtained (g_0 = 1.911; $g_{\parallel}$ = 1.993; $g_{\perp}$ = 1.869) agree with values calculated from the ultraviolet/visible transitions and the measured magnetic moment.[10,11] Infrared and electronic absorption spectra are given in Table II. Although it is thermally unstable, the compound gives a mass spectrum containing the parent molecular ion.[9] The crystalline complex has the same trigonal structure as the Fe compound.[14]

C. TRIS[BIS(TRIMETHYLSILYL)AMIDO]VANADIUM(III) (Tris(1,1,1,3,3,3-hexamethyldisilazanato)vanadium(III))

$$VCl_3 + 2N(CH_3)_3 \longrightarrow VCl_3[N(CH_3)_3]_2$$

$$VCl_3[N(CH_3)_3]_2 + 3LiN[Si(CH_3)_3]_2 \longrightarrow V\{N[Si(CH_3)_3]_2\}_3 + 3LiCl + 2N(CH_3)_3$$

Procedure

Trichlorobis(trimethylamine)vanadium (7.16 g, 0.026 mole), prepared in sealed tubes,[8] is added slowly by means of a transfer tube to a cooled sodium-dried benzene solution of 12.86 g (0.077 mole) of lithium bis(trimethylsilyl)amide (total volume about 120 mL) contained in a 250-mL, round-bottomed flask. A dark-red solution forms. This is stirred at room temperature for 48 hours and the white precipitate is filtered in the usual way. The filtrate is concentrated to about a third of its original bulk and stored under nitrogen at 0° for 24 hours, by which time small brown needles have crystallized. These are filtered off rapidly, washed with a small amount of cold benzene, dried under vacuum, and transferred into Pyrex ampules, which are then sealed. The yield of tris[bis(trimethylsilyl)amido]vanadium(III) is about 5 g (35%). *Anal.* Calcd. for $V\{N[Si(CH_3)_3]_2\}_3$: V, 9.57; C, 40.63; H, 10.23, N, 7.90. Found: V, 9.52; C, 40.41; H, 10.01; N, 7.74

Properties

Tris[bis(trimethylsilyl)amido]vanadium(III) crystallizes from benzene as dark-brown, soft needles, extremely sensitive to air and moisture. This complex is paramagnetic ($\mu_{eff} \sim$ 2.4 BM) but does not show electron paramagnetic resonance absorption at temperatures above that of liquid nitrogen. The compound is thermally unstable but gives a mass spectrum containing the parent molecular ion. Infrared spectra and electronic absorption spectra are given in Table II. The crystalline complex has the same trigonal structure as the Fe compound.[14]

D. TRIS[BIS(TRIMETHYLSILYL)AMIDO]CHROMIUM(III) (Tris(1,1,1,3,3,3-hexamethylsilazanato)chromium(III))

$$CrCl_3 + 3LiN[Si(CH_3)_3]_2 \longrightarrow Cr\{N[Si(CH_3)_3]_2\}_3 + 3LiCl$$

Procedure

Freshly distilled bis(trimethylsilyl)amine (20 mL, 0.10 mole) contained in a 50 mL dropping funnel is added dropwise to 60 mL (0.093 mole) of the standard butyllithium solution. The experimental arrangement and solvent are exactly the same as for synthesis A except that anhydrous chromium trichloride (sublimed) is added by means of the transfer tube (9.6 g, 0.06 mole). The solution turns green and a white precipitate forms after a few hours. Stirring is continued for 24 hours, the mixture is brought to reflux briefly with an infrared lamp, and the solvent is stripped. Pumping on the dry residue is continued for 6 hours. The green residue is extracted with 100 mL of dry pentane as in synthesis

A, the precipitate is filtered, and the solvent volume is reduced to about 60 mL to give a large crop of bright green crystals. The yield of tris[bis(trimethylsilyl)-amido]chromium(III) is about 12 g (75%). *Anal.* Calcd. for $Cr\{N[Si(CH_3)_3]_2\}_3$: C, 40.55; H, 10.21; N, 7.88 Found: C, 40.42; H, 10.13; N, 7.74.

Properties

Tris[bis(trimethylsilyl)amido]chromium(III) crystallizes from pentane as apple-green needles, sensitive to air and moisture. The compound sublimes at 110° at 0.005 torr and gives a mass spectrum containing the parent molecule ion.[9] Magnetic measurements show that the metal atom has three unpaired electrons,[9] and EPR studies reveal a large zero field splitting.[10,11] Infrared spectra and electronic absorption spectra are given in Table II. Its crystal structure is the same as that of the Fe compound.[14] The compound reacts with nitric oxide in pentane and gives a stable brown diamagnetic mononitrosyl derivative $Cr(NO)\{N[Si(CH_3)_3]_2\}_3$.[12]

E. TRIS[BIS(TRIMETHYLSILYL)AMIDO]IRON(III) (Tris(1,1,1,3,3,3-hexamethylsilazanato)iron(III))

$$FeCl_3 + 3LiN[Si(CH_3)_3]_2 \longrightarrow Fe\{N[Si(CH_3)_3]_2\}_3 + 3LiCl$$

Procedure

The method is essentially the same as that described in synthesis A. The quantities of bis(trimethylsilyl)amine (0.10 mole) and butyllithium (0.093 mole) are again used, but the solvent is 100 mL of dry benzene (sodium-dried and distilled from calcium hydride). The addition of the anhydrous $FeCl_3$ (9.8 g, 0.06 mole) leads to a slight warming of the reaction mixture. After 24 hours of stirring at room temperature, by which time the solution has assumed a dark-green coloration, the contents are briefly heated to reflux with an infrared lamp and the mixture is filtered. Concentration of the filtrate produces a mass of dark-green needle crystals, which are filtered, washed with a small amount of cold benzene, pumped dry, and sealed into ampules under vacuum. The yield of tris-[bis(trimethylsilyl)amido]iron(III) is about 10 g (62%). Satisfactory analyses have not been obtained due to this compound's high reactivity to air and moisture.

Properties

Tris[bis(trimethylsilyl)amido]iron(III) crystallizes from benzene as dark-green, almost black, needles, very sensitive to air and moisture. The compound sublimes at 120° at 0.005 torr and gives a mass spectrum containing the parent molecular ion.[9] Magnetic measurements show that the metal atom has five unpaired electrons, and single crystal EPR measurements[11] and Mössbauer spectra[15] have been reported. Infrared spectra and electronic absorption spectra are given in Table II.

References

1. H. Bürger and U. Wannagat, *Monatsh. Chem.*, **94**, 1007 (1963).
2. H. Bürger and U. Wannagat, *Monatsh. Chem.*, **95**, 1099 (1964).
3. R. G. Copperthwaite, Ph.D. Thesis, London, 1971.
4. D. F. Shriver, *The Manipulation of Air-Sensitive Compounds,* McGraw-Hill Book Co., New York, 1969.
5. R. J. H. Clark, *Inorg. Synth.*, **13**, 166 (1972).
6. R. C. Osthoff and S. W. Kantor, *Inorg. Synth.*, **5**, 58 (1957).
7. E. H. Amonoo-Naizer, R. A. Shaw, D. O. Skovlin, and B. C. Smith, *Inorg. Synth.*, **13**, 19 (1972).
8. A. T. Casey, R. J. H. Clark, and K. J. Pidgeon, *Inorg. Synth.*, **13**, 179 (1972).
9. E. C. Alyea, D. C. Bradley, and R. G. Copperthwaite, *J. Chem. Soc. (Dalton),* **1972**, 1580.
10. E. C. Alyea, D. C. Bradley, R. G. Copperthwaite, and K. D. Sales, *J. Chem. Soc. (Dalton),* **1973**, 185.
11. D. C. Bradley, R. G. Copperthwaite, S. A. Cotton, K. D. Sales, and J. F. Gibson, *J. Chem. Soc. (Dalton),* **1973**, 191.
12. D. C. Bradley and C. W. Newing, *Chem. Commun.*, **1970**, 219.
13. J. S. Ghotra, M. B. Hursthouse, and A. J. Welch, *Chem. Commun.*, **1973**, 669.
14. D. C. Bradley, M. B. Hursthouse, and P. F. Rodesiler, *Chem. Commun.*, **1969**, 14; M. B. Hursthouse and P. F. Rodesiler, *J. Chem. Soc. (Dalton),* **1972**, 2100; C. E. Heath and M. B. Hursthouse, to be published.
15. B. W. Fitzsimmons and C. E. Johnson, *Chem. Phys. Lett.*, **24**, 422 (1974).

19. BIS(TRIPHENYLPHOSPHINE)PLATINUM COMPLEXES

Submitted by D. M. BLAKE* and D. M. ROUNDHILL†
Checked by C. AMBRIDGE,‡ S. DWIGHT,‡ and H. C. CLARK‡

A. CARBONATOBIS(TRIPHENYLPHOSPHINE)PLATINUM(II)

$$Pt[P(C_6H_5)_3]_4 + O_2 + CO_2 \xrightarrow{C_6H_6} Pt[P(C_6H_5)_3]_2(CO_4) + 2(C_6H_5)_3P$$

*Department of Chemistry, University of Texas at Arlington, Arlington, TX 76019.
†Department of Chemistry Washington State University, Pullman, WA 99163.
‡Department of Chemistry, The University of Western Ontario, London, Canada.

$$Pt[P(C_6H_5)_3]_2CO_4 + (C_6H_5)_3P \xrightarrow[C_6H_6]{CH_2Cl_2}$$

$$[Pt[P(C_6H_5)_3]_2(CO_3)] \cdot C_6H_6 + OP(C_6H_5)_3$$

Carbonatobis(triphenylphosphine)platinum(II)[1] is useful for the preparation of dianionobis(triphenylphosphine)platinum(II) complexes[2] as well as olefin,[3] acetylene[3] and carbonyl[3] derivatives of platinum(0). The method can be used to prepare $Pt[P(C_6H_5)_2CH_3]_2(CO_3)$[3] and $Pt[As(C_6H_5)_3]_2(CO_3)$.[1] A procedure is available[4] for the preparation of *cis*-[diacetatobis(diphenylphosphino)platinum] (*cis*-$[Pt(CH_3CO_2)_2[P(C_6H_5)_2]_2]$).

Procedure

Carbon dioxide and oxygen are bubbled through a solution of 5 g of tetrakis(triphenylphosphine)platinum, $Pt[P(C_6H_5)_3]_4$,[5] (4 mmole) in 120 mL of benzene. After 30 minutes the mixture is very pale yellow. The solid is recovered using a medium-porosity filter and is washed with benzene (50 mL). The crude product (3-3.5 g) is a mixture of the carbonato and peroxycarbonato complexes. To complete the conversion to the carbonato complex, the crude product is dissolved in 75 mL of dichloromethane, and 5 g (19 mmole) of triphenylphosphine is added. The mixture is refluxed overnight. Benzene (75 mL) is added to the mixture and the volume is reduced by half on a rotary evaporator. The colorless solid is recovered using a medium-porosity filter, washed with benzene (50 mL) and diethyl ether (25 mL), and then dried *in vacuo*. The yield is 2.5 g (73%). *Anal.* (as benzene solvate) Calcd. for $PtC_{43}H_{36}P_2O_3$: C, 60.2; H, 4.2. Found: C 59.6; H, 4.1.

Properties

The compound is an air-stable white solid. Prolonged exposure to room light results in some decomposition. It is soluble in chloroform and dichloromethane and insoluble in alcohols, benzene, and diethyl ether. The peroxycarbonato complex may be detected, if present as an impurity, by an infrared band at 780 (m) cm^{-1} assigned to $\nu(O-O)$.[1] The infrared spectrum of the carbonato complex shows $\nu(C=O)$ at 1685 (vs) cm^{-1} mp 202-205° (dec.).

B. (ETHYLENE)BIS(TRIPHENYLPHOSPHINE)PLATINUM(O)

$$Pt[P(C_6H_5)_3]_2(CO_3) + 2NaBH_4 + C_2H_4 \xrightarrow{C_2H_5OH}$$

$$Pt[P(C_6H_5)_3]_2(C_2H_4) + Na_2CO_3 + H_2 + B_2H_6{}^*$$

*Diborane is not detected and presumably is consumed rapidly in the reaction mixture.

The preparation is based on a convenient starting compound that may be stored readily. The procedure can be used to prepare adducts of other olefins and acetylenes. The ethylene complex is of widespread use in the study of oxidative addition reactions of platinum(0).[6-8]

Procedure

■ **Caution.** *Because of the toxicity and flammability of ethylene, the reaction should be carried out in an efficient hood.*

A suspension of 1.5 g (1.75 mmole) of $[Pt[P(C_6H_5)_3]_2(CO_3)] \cdot C_6H_6$ (Sec. A) in 40 mL of ethanol is placed in a 100-mL, two-necked flask. An addition funnel is fitted to one neck and a tube delivering a slow stream of ethylene below the surface of the mixture is connected by way of the other neck. The mixture is stirred rapidly as 20 mL of 0.1 *M* sodium tetrahydroborate(1−), $NaBH_4$, in ethanol is added dropwise over a period of 20 minutes. The white suspension is stirred for 1 hour, during which time the flow of ethylene is maintained. The solid is recovered by filtration using a medium-porosity filter, washed with ethanol (20 mL), water (25 mL), and finally ethanol (25 mL), and then dried *in vacuo*. The yield is 1.2 g (90%). *Anal.* Calcd. for $PtC_{38}H_{34}P_2$: C, 61.03; H, 4.59. Found: C, 60.76; H, 4.68.

Properties

The compound is a white solid that may be handled in air for short periods without apparent decomposition and can be stored under nitrogen indefinitely. It is soluble in benzene and dichloromethane and insoluble in alcohols. The mp is 126-128° (dec.); the checkers observed 122-126°.

C. (DIPHENYLACETYLENE)BIS(TRIPHENYLPHOSPHINE)-PLATINUM(0)[9-11]

$$Pt[P(C_6H_5)_3]_2(CO_3) + C_6H_5C{\equiv}CC_6H_5 \xrightarrow{C_2H_5OH}$$
$$Pt[P(C_6H_5)_3]_2(C_6H_5C{\equiv}CC_6H_5) + CO_2 + C_2H_4O + H_2O$$

Method 1

(Diphenylacetylene)bis(triphenylphosphine)platinum(0) is useful for the preparation of anionobis(triphenylphosphine)vinylplatinum(II) complexes,[12,13] as well as the corresponding aniono derivatives.[14] The method can be used to prepare analogous complexes covering a wide range of acetylenes. A similar procedure has been used to prepare silylplatinum complexes from organosilicon hydrides and carbonatobis(*tert*-phosphine)platinum(II) complexes.[15]

Procedure

A mixture of 0.5 g (0.6 mmole) of $[Pt[P(C_6H_5)_3]_2(CO_3)] \cdot C_6H_6$ and 0.5 g (2.8 mmole) diphenylacetylene is refluxed in 50 mL of ethanol for 4 hours. The mixture is chilled in ice and filtered through a medium-porosity filter. The light-cream-colored product is washed with ethanol (15 mL) and dried *in vacuo*. The yield is 0.52 g (96%). *Anal.* Calcd. for $PtC_{50}H_{40}P_2$: C, 66.9; H, 4.5. Found: C, 66.9; H, 4.5.

$$2PtCl_2[P(C_6H_5)_3]_2 + 2C_6H_5C{\equiv}CC_6H_5 + 3N_2H_4 \longrightarrow 2Pt[P(C_6H_5)_3]_2(C_6H_5C{\equiv}CC_6H_5) + N_2 + 4HCl$$

$$2N_2H_4 + 4HCl \longrightarrow 2[N_2H_6]Cl_2$$

Method 2

Procedure

Under a nitrogen atmosphere, 0.4 mL of an 85% hydrazine solution is added dropwise to a stirred suspension of 0.47 g (0.6 mmole) of $PtCl_2[P(C_6H_5)_3]_2$[16] in 25 mL of nitrogen-purged ethanol. The reaction mixture becomes clear yellow over 2-3 minutes. To this mixture a solution of 0.5 g (2.8 mmole) of diphenylacetylene dissolved in 15 mL of ethanol is added. The mixture is heated to reflux and then allowed to cool slowly to room temperature. After 30 minutes the product is collected on a medium-porosity filter. The compound is washed with 10 mL of water-methanol (1:9) solution and dried *in vacuo*. The yield is 0.44 g (83%). The checkers obtained 91-95% yields using half the scale given above.

Properties

The compound is an air-stable solid. It is soluble in chloroform, dichloromethane, and benzene but insoluble in alcohol and diethyl ether. The infrared spectrum of the compound shows $\nu(C{\equiv}C)$ at 1740 (s) cm^{-1}. The melting point is 160-165° (dec.).

References

1. P. J. Hayward, D. M. Blake, G. Wilkinson, and C. J. Nyman, *J. Am. Chem. Soc.*, **92**, 5873 (1970).

2. C. J. Nyman, C. E. Wymore, and G. Wilkinson, *J. Chem. Soc., A*, **1968**, 561.
3. D. M. Blake and L. M. Leung, *Inorg. Chem.*, **11**, 2879 (1972).
4. A. Dobson, S. D. Robinson, and M. F. Uttley, *Inorg. Synth.*, **17**, 124 (1977).
5. R. Ugo, F. Cariati, and G. La Monica, *Inorg. Synth.*, **11**, 105 (1968).
6. J. P. Birk, J. Halpern, and A. L. Pickard, *J. Am. Chem. Soc.*, **90**, 4491 (1968).
7. C. D. Cook and G. S. Jauhal, *J. Am. Chem. Soc.*, **90**, 1464 (1968).
8. D. M. Blake, S. Shields, and L. Wyman, *Inorg. Chem.*, **13**, 1595 (1974).
9. J. Chatt, G. A. Rowe, and A. A. Williams, *Proc. Chem. Soc.*, **1957**, 208.
10. J. H. Nelson, H. B. Jonassen, and D. M. Roundhill, *Inorg. Chem.*, **8**, 2591 (1969).
11. A. D. Allen and C. D. Cook, *Can. J. Chem.*, **42**, 1063 (1964).
12. P. B. Tripathy, B. W. Renoe, K. Adzamli, and D. M. Roundhill, *J. Am. Chem. Soc.*, **93**, 4406 (1971).
13. B. E. Mann, B. L. Shaw, and N. I. Tucker, *J. Chem. Soc. A*, **1971**, 2667.
14. P. B. Tripathy and D. M. Roundhill, *J. Organometal. Chem.*, **24**, 247 (1970).
15. C. Eaborn, T. N. Metham, and A. Pidcock, *J. Chem. Soc. (Dalton)*, **1975**, 2212.
16. J. C. Bailar, Jr., and H. Itatani, *Inorg. Chem.*, **4**, 1618 (1965).

20. SULFUR NITRIDE COMPLEXES OF NICKEL

$$S_4N_4 + NiCl_2 \xrightarrow{CH_3OH} Ni(S_2N_2H)_2 + (S_3N)Ni(S_2N_2H) + Ni(S_3N)_2$$

Submitted by D. T. HAWORTH,* J. D. BROWN,* and Y. CHEN*
Checked by WILLIAM L. HOLDER† and WILLIAM L. JOLLY†

Thionitrosyl (NS) was unknown as a monodentate ligand until the recent preparation of a molybdenum thionitrosyl complex by Chatt and Dilworth.[1] However, there are a number of examples of metal chelates containing the bidentate ligands, S_2N_2 or S_2N_2H, the first of which was $Pb(S_2N_2)$ prepared in 1902 by Ruff and Geisel.[2] The syntheses of nickel complexes containing S_3N and S_2N_2H as bidentate ligands are described below.[3]

Procedure

■ **Caution.** *This synthesis involves the use of S_4N_4, which is a mild explosive. It can explode on rubbing or scraping, for example, in the neck of a bottle with a ground-glass stopper.*‡

*Marquette University, Milwaukee, WI 53233.
†Department of Chemistry, University of California, Berkeley, CA 94720.
‡See *Inorg. Synth.*, **17**, 197 (1977).

A 250-mL, two-necked flask is charged with 3.0 g (0.0163 mole) of S_4N_4,[4] 3.75 g (0.0289 mole) of anhydrous $NiCl_2$,[5] and 120 mL of anhydrous methanol. Under a slow stream of nitrogen, the flask is heated to reflux for 8 hours, after which the alcohol is distilled under vacuum on a rotary evaporator at ambient temperature. The remaining dry solid is extracted with 200 mL of dry benzene. The benzene solution is put on a chromatographic column (20 mm diameter) containing 100 g of acid-washed alumina. The absorbent is prepared by pipetting 3 mL of distilled water in a dry flask, which on rotation evenly distributes the water on the surface of the glass. The aluminum oxide is then added and the rotation is continued until no lumps or moist spots can be detected. This mixture is allowed to stand for 1-2 hours.

The first fraction eluted is green, the second fraction is red, and a third purple-colored fraction remains on the column. The solid remaining in the flask is extracted further with two 100-mL portions of acetone. The acetone portions, when put on the column, slowly elute the purple fraction. The purple fraction is eluted completely by an ethanol/acetone (1:4) solution.

Since a clean separation of the red and purple fractions may not be achieved on the first chromatogram the above procedure is repeated on the black solid obtained from evaporation of the purple fraction, which contains $Ni(S_2N_2H)_2$. The black crystals (2.8 g) obtained from the second chromatogram can be purified further by crystallization from acetone by slow addition of pentane and finally by sublimation of these crystals *in vacuo* at 140°. The yield of the product is 71% (based on S_4N_4); mp 154-155° (literature, mp 153°[7], 154.7-155°[3]. The green fraction contains $Ni(S_3N)_2$ and the red fraction contains the mixed ligand complex $Ni(S_2N_2H)(S_3N)$. These latter compounds are obtained in very low yields. A scale-up of the original reaction mixture is necessary to obtain these latter compounds in milligram quantities. The red compound, $Ni(S_2N_2H)$-(S_3H), if chromatographed on new acid-washed alumina may be disproportionated partially to $Ni(S_2N_2H)_2$ and $Ni(S_3N)_2$. *Anal.* Calcd. for $H_2N_4S_4Ni$: H, 0.82; N, 22.87, S, 52.3. Found: H, 0.89; N, 22.81; S, 52.0.

Properties

The molecule $Ni(S_2N_2H)_2$ has a planar geometry in which the N—H protons are in a cis arrangement.[6] The compound is a black crystalline solid and is soluble in acetone, slightly soluble in benzene, and insoluble in pentane. The benzene solution is green to reflected light and purple to transmitted light. The infrared spectrum obtained using the KBr pellet technique shows major absorptions bands at 3283 (s), 3162 (s), 1260 (w), 1166 (m), 1039 (vs), 890 (s), 719 (vs), 702 (vs), 626 (m), 588 (w) and 524 (m) cm^{-1}. Its low-frequency spectrum in a Nujol mull on a polyethylene window gave absorption bands at 395 (w), 356 (w), 331 (s), 321 (m), and 314 (m) cm^{-1}. The ^1H NMR spectrum of this compound in

$(CD_3)_2CO$ shows a broad band at −9.8 ppm (N−H), using TMS as an internal standard. The five largest *d* values (Å) for its X-ray powder pattern are in decreasing order of intensity: 3.32, 6.17, 1.66, 4.83, and 5.64.

The deuterated complex can be prepared by dissolving $Ni(S_2N_2H)_2$ in tetrahydrofuran and adding a 1000-fold excess of D_2O. The mixture is stirred at ambient temperature for 3 hours. The solvent is removed at reduced pressure on a rotary evaporator. The infrared spectrum shows shifts of the 3283 and 3162 cm^{-1} bands in $Ni(S_2N_2H)_2$ to 2433 and 2375 cm^{-1}, respectively, in $Ni(S_2N_2D)_2$. These are assigned to the N−H and N−D antisymmetric and symmetric stretchin modes. The 1166 cm^{-1} band also shows a large shift (182 cm^{-1}) on deuteration.

This synthesis can be extended to the preparation of the sulfur nitride complexes of other transition and post-transition metals.[7-11] Furthermore, $Ni(S_2N_2H)_2$ can serve as a starting material for the synthesis of many derivatives of this compound by replacement of the acidic N−H proton.[12] Such compounds include substituents on one or two of the nitrogen atoms, as well as groups that bridge the *cis*-nitrogen atoms.

References

1. J. Chatt and J. R. Dilworth, *Chem. Commun.*, **13**, 508 (1974).
2. O. Ruff and E. Geisel, *Ber. Deut. Chem. Ges.*, **37**, 1573 (1902).
3. T. S. Piper, *J. Am. Chem. Soc.*, **80**, 30 (1958).
4. M. Villena and W. L. Jolly, *Inorg. Synth.*, 9, 98 (1967).
5. A. R. Pray, *Inorg. Synth.*, **5**, 153 (1957).
6. J. Weiss and U. Thewalt, *Z. Anorg. Allgem. Chem.*, **363**, 159 (1968).
7. M. Goehring and A. Debo, *Z. Anorg. Allgem. Chem.*, **273**, 319 (1953).
8. M. Goehring, K. W. Kaum, and J. W. Weiss, *Z. Naturforsch.*, **106**, 298 (1955).
9. K. W. Daum, M. Goehring, and J. Weiss, *Z. Anorg. Allgem. Chem.*, **278**, 260 (1955).
10. M. Goehring and K. W. Daum, *ibid.*, **282**, 83 (1955).
11. E. Fluck, M. Goehring, and J. Weiss, *ibid.*, **287**, 51 (1956).
12. J. Weiss, *Fortschr. Chem. Forsch.*, **5**, 635 (1966).

21. (η^5-CYCLOPENTADIENYL)NITROSYL COMPLEXES OF CHROMIUM, MOLYBDENUM, AND TUNGSTEN

Submitted by JAMES K. HOYANO,* PETER LEGZDINS,* and JOHN T. MALITO*
Checked by THOMAS ARNOLD† and BASIL I. SWANSON†

*Department of Chemistry, the University of British Columbia, Vancouver, B. C., Canada, V6T 1W5.

†Department of Chemistry, The University of Texas at Austin, Austin, TX 78712.

The chemistry of the dicarbonyl(η^5-cyclopentadienyl)nitrosyl and the chloro(η^5-cyclopentadienyl)dinitrosyl complexes of chromium, molybdenum, and tungsten [i.e., [(η^5-C_5H_5)M(CO)$_2$(NO)] and [(η^5-C_5H_5)M(NO)$_2$Cl] has not been studied extensively, partly because of the various difficulties associated with their preparation.[1] The procedures described below are of general applicability to all three metals and lead to the desired compounds in high yields. The carbonyl nitrosyl complexes are the synthetic precursors of the chloro nitrosyl complexes and so their preparation is described first.

A. DICARBONYL(η^5-CYCLOPENTADIENYL)NITROSYL COMPLEXES OF CHROMIUM, MOLYBDENUM, AND TUNGSTEN

$$NaC_5H_5 + M(CO)_6 \longrightarrow Na[(\eta^5\text{-}C_5H_5)M(CO)_3] + 3CO \qquad (1)$$

$$Na[(\eta^5\text{-}C_5H_5)M(CO)_3] + p\text{-}CH_3C_6H_4SO_2N(NO)CH_3 \xrightarrow{THF} (\eta^5\text{-}C_5H_5)M(CO)_2(NO) + CO + p\text{-}CH_3C_6H_4SO_2N(CH_3)Na \qquad (2)$$

Dicarbonyl(η^5-cyclopentadienyl)nitrosylchromium may be prepared in good yield by the action of nitric oxide on [(η^5-C_5H_5)Cr(CO)$_3$]$_2$,[2] but the latter reagent can be obtained only in low yields and with the expenditure of much effort.[3] The analagous carbonyl nitrosyl complexes of molybdenum and tungsten are formed in low yields when aqueous solutions of Na[(η^5-C_5H_5)M(CO)$_3$] (M = Mo or W) are treated with nitric oxide.[4] A more general nitrosylating agent is *N*-methyl-*N*-nitroso-*p*-toluenesulfonamide (Diazald), which converts both (η^5-C_5H_5)M(CO)$_3$H (M = Mo or W)[5-7] and [(η^5-C_5H_5)M(CO)$_3$]$^-$ (M = Cr, Mo, or W)[8] into the desired (η^5-C_5H_5)M(CO)$_2$(NO) compounds. To effect the above reactions in high yield, it is of paramount importance that the Na[(η^5-C_5H_5)M(CO)$_3$] salts prepared in reaction 1 do not contain a large excess of unreacted Na[C_5H_5]. The molybdenum and tungsten salts are obtained according to the published procedure,[5] although refluxing for 3 days is advised for the reaction of the tungsten compound to ensure complete conversion. The chromium analogue is best prepared in the follow manner.

Procedure

■ **Caution.** *Unless otherwise indicated, this and all other reactions and manipulations are carried out under nitrogen in a well-ventilated fume hood.*

A 200-mL, three-necked flask is fitted with a nitrogen inlet and stirrer and is thoroughly flushed with prepurified nitrogen. Into the flask is syringed a tetrahydrofuran (THF) solution containing 4.18 g (47.5 mmole) of NaC_5H_5.[3] The THF is removed *in vacuo* and 11.00 g (50.0 mmole) of $Cr(CO)_6$ (Pressure

Chemical Co., Pittsburgh, PA 15201) and 100 mL of di-*n*-butyl ether [Eastman Kodak practical grade dried by distillation from calcium hydride (CaH_2) and deaerated with nitrogen] are added. The flask is then equipped with a Liebig condenser and the reaction mixture is refluxed with vigorous stirring for 12 hours. During this time the reaction vessel is shaken occasionally to reintroduce any sublimed $Cr(CO)_6$ into the refluxing reaction mixture. The final reaction mixture is allowed to cool to room temperature and filtered, and the pale-yellow solid thus collected is washed with di-*n*-butyl ether (3 × 10 mL) and dried under nitrogen. The excess $Cr(CO)_6$ and any di-*n*-butyl ether remaining in this solid are removed by sublimation at 90°/0.005 torr onto a water-cooled probe. The $Na[(\eta^5\text{-}C_5H_5)M(CO)_3]$ complexes of molybdenum and tungsten[5] are freed of any unreacted hexacarbonyl in a similar manner, and all three sodium salts are used without further purification.

The preparations of all three $(\eta^5\text{-}C_5H_5)M(CO)_2(NO)$ complexes from their corresponding $[(\eta^5\text{-}C_5H_5)M(CO)_3]^-$ anions are similar. The experimental procedure, using the tungsten complex (dicarbonyl(η^5-cyclopentadienyl)nitrosyltungsten) as a typical example, is given below.

A 300-mL, three-necked flask is equipped with a nitrogen inlet, an addition funnel, and a stirrer. It is charged with 17.3 g (48.5 mmole) of $Na[(\eta^5\text{-}C_5H_5)W(CO)_3]$ and 120 mL of THF (Fisher Scientific Co. reagent grade dried by distillation from lithium tetrahydridoaluminate ($LiAlH_4$) and deaerated with nitrogen). A THF solution (50 mL) containing 10.4 g (48.6 mmole) of Diazald (*N*-methyl-*N*-nitroso-*p*-toluenesulfonamide, Eastman Kodak reagent grade) is syringed into the addition funnel. The solution of Diazald is added dropwise over a period of 15 minutes to the stirred reaction mixture. Gas evolution occurs and an orange-brown solid precipitates. The mixture is stirred for an additional 15 minutes and the solvent is removed *in vacuo.* Sublimation of the resulting brown residue at 50-60°/0.005 torr onto a water-cooled probe for 3 days affords 13.6 g (84% yield) of $(\eta^5\text{-}C_5H_5)W(CO)_2(NO)$. The corresponding chromium and molybdenum complexes are obtained similarly in yields of 60 and 93%, respectively.

Anal. Calcd. for $C_5H_5Cr(CO)_2(NO)$: C, 41.39; H, 2.48; N, 6.90. Found: C, 41.40; H, 2.60; N, 6.70. $\nu_{CO}(CH_2Cl_2)$ cm^{-1}; 2020 (s); 1945 (s). $\nu_{NO}(CH_2Cl_2)$ cm^{-1}: 1680 (s).

Calcd. for $C_5H_5Mo(CO)_2(NO)$: C, 34.03; H, 2.04; N, 5.67. Found: C, 34.28, H, 2.24; N, 5.54. $\nu_{CO}(CH_2Cl_2)$ cm^{-1}: 2020 (s); 1937 (s). $\nu_{NO}(CH_2Cl_2)$ cm^{-1}: 1663 (s).

Calcd. for $C_5H_5W(CO)_2(NO)$: C, 25.10; H, 1.50; N, 4.18. Found: C, 25.29; H, 1.70; N, 4.13. $\nu_{CO}(CH_2Cl_2)$ cm^{-1}: 2010 (s), 1925 (s). $\nu_{NO}(CH_2Cl_2)$ cm^{-1}: 1655 (s).

Properties

The properties have been discussed previously.[1,9] The compounds are orange to orange-red solids readily soluble in organic solvents. The solids are stable in air for short periods of time and indefinitely under nitrogen.

B. CHLORO(η^5-CYCLOPENTADIENYL)DINITROSYL COMPLEXES OF CHROMIUM, MOLYBDENUM, AND TUNGSTEN

$$(\eta^5\text{-}C_5H_5)M(CO)_2(NO) + ClNO \xrightarrow{CH_2Cl_2} (\eta^5\text{-}C_5H_5)M(NO)_2Cl + 2CO$$

$$M = Cr, Mo, \text{ or } W$$

Chloro(η^5-cyclopentadienyl)dinitrosylchromium, $(\eta^5\text{-}C_5H_5)Cr(NO)_2Cl$, may be prepared by the reaction of nitric oxide with $[(\eta^5\text{-}C_5H_5)CrCl_2]_2$.[5,10,11] The molybdenum compound is obtained in low yields by the reaction of cyclopentadienylthallium (TlC_5H_5) with $[Mo(NO)_2Cl_2]_n$[12] and by the reaction of sodium nitrate with $[(\eta^5\text{-}C_5H_5)Mo(CO)_3(NH_3)]Cl$ in hydrochloric acid.[13] The tungsten analogue can be prepared by the treatment of $[(\eta^5\text{-}C_5H_5)W(NO)_2(CO)][PF_6]$ with sodium chloride.[14] The general method of synthesis described below involves the reaction of nitrosyl chloride with the dicarbonyl (η^5-cyclopentadienyl)nitrosyl complexes.

Procedure

■ **Caution.** *All reactions and manipulations are preformed under nitrogen in a well-ventilated fume hood.*

Approximately 2-3 mL of nitrosyl chloride (either freshly prepared[15] or purchased from Matheson of Canada, Whitby, Ontario) is condensed into a 5-mL graduated cold trap held at -78°. It is then allowed to melt so that its volume can be measured, and it is distilled under static vacuum into a 100-mL, two-necked flask held at -78°. Approximately 30-40 mL of dichloromethane [Fisher Scientific Co. reagent grade freshly distilled from phosphorus oxide (P_2O_5) and deaerated with nitrogen] is syringed into this flask, and the resulting red solution is stored at -78° until just prior to use when it is allowed to warm to room temperature.

All three $(\eta^5C_5H_5)M(NO)_2Cl$ complexes are prepared in a similar manner, but to achieve maximum yields of the chromium and tungsten compounds, the reactions must be performed at -78°. The molybdenum complex (chloro(η^5-cyclopentadienyl)nitrosylmolybdenum), on the other hand, can be obtained in excellent yields even at room temperature and its preparation is detailed below.

A 200-mL, three-necked flask, equipped with a stirrer, a nitrogen inlet, and an addition funnel, is charged with 6.5 g (26 mmole) of $(\eta^5\text{-}C_5H_5)Mo(CO)_2(NO)$ and 100 mL of dichloromethane. The nitrosyl chloride solution (typically containing 2.3 mL (50 mmole) of nitrosyl chloride in 30 mL of dichloromethane) is added dropwise to the stirred reaction mixture. Gas evolution occurs and the orange solution becomes dark green. The reaction is monitored by infrared spectroscopy and the nitrosyl chloride solution is added until the carbonyl absorptions of the initial reactant have disappeared. It is extremely important that the stoichiometric amount of nitrosyl chloride be used, since even a slight excess of ClNO reduces significantly the yields of the desired products, especially in the cases of the chromium and tungsten complexes. Conversely, if an insufficient quantity of ClNO is added, the separation of any unreacted (η^5-C_5H_5)M(CO)$_2$(NO) from the desired chloronitrosyl complex is difficult. The final reaction mixture is concentrated *in vacuo* to approximately 30 mL and is filtered through a short (3 X 5 cm) Floricil column. The column is washed with dichloromethane until the washings are colorless, and the combined filtrates are then concentrated *in vacuo* to a volume of ~30 mL. Hexane [Fisher Scientific Co. reagent grade dried by distillation from lithium tetrahydridoaluminate ($LiAlH_4$) and deaerated with nitrogen] is added until crystallization appears to be complete; approximately 100-125 mL of hexane is required. The resuling green crystals are collected by filtration, washed with hexane (2 X 15 mL), and dried under nitrogen to obtain 6.0 g (89% yield) of analytically pure (η^5-C_5H_5)-Mo(NO)$_2$Cl. The chromium and tungsten compounds are obtained similarly (except that the reaction flask is maintained at −78° during the addition of the ClNO solution) in yields of 77 and 72%, respectively.

Anal. Calcd. for C_5H_5Cr(NO)$_2$Cl: C, 28.26; H, 2.37; N, 13.18; Cl, 16.68. Found. C, 28.29; H, 2.55; N, 12.86; Cl, 16.99. ν_{NO}(CH_2Cl_2) cm^{-1}: 1816 (s); 1711 (s). mp 144° (dec).

Calcd. for C_5H_5Mo(NO)$_2$Cl: C, 23.42; H, 1.96; N, 10.92. Found: C, 23.53; H, 1.90; N, 10.70. ν_{NO}(CH_2Cl_2) cm^{-1}: 1759 (s); 1665 (s). mp 116°

Calcd. for C_5H_5W(NO)$_2$Cl: C, 17.44; H, 1.46; N, 8.13; Cl, 10.29. Found: C, 17.68; H, 1.62 N, 8.10; Cl, 10.26. ν_{NO}(CH_2Cl_2) cm^{-1}: 1733 (s); 1650 (s). mp 127° (dec).

Properties

Chloro(η^5-cyclopentadienyl)dinitrosylchromium is a gold crystalline solid that is slightly soluble in hexane but freely soluble in benzene, tetrahydrofuran, and dichloromethane. The molybdenum and tungsten analogues are green crystalline solids with similar solubility properties. All three compounds are quite stable under nitrogen at room temperature and may be exposed to air for short periods of time without noticeable decomposition. The compounds may be sublimed at 40-50°/0.005 torr, although some attendant decomposition occurs with the chromium and tungsten species. The PMR spectra (benzene-d_6 solutions) exhibit sharp resonances at τ5.22, τ4.93 and τ5.02 due to the cyclopentadienyl protons of the chromium, molybdenum and tungsten complexes, respectively. Chemically the $(\eta^5\text{-}C_5H_5)M(NO)_2Cl$ species are useful precursors for the synthesis of a wide variety of thermally stable alkyl and aryl complexes, $(\eta^5\text{-}C_5H_5)M(NO)_2R$.

References

1. K. W. Barnett and D. W. Slocum, *J. Organometal. Chem.*, **44**, 1 (1972).
2. E. O. Fischer and K. Plesske, *Chem. Ber.*, **94**, 93 (1961).
3. R. B. King and F. G. A. Stone, *Inorg. Synth.*, 7, 99 (1963).
4. E. O. Fischer, O. Beckert, W. Hafner, and H. O. Stahl, *Z. Naturforsch.*, **10b**, 598 (1955).
5. T. S. Piper and G. Wilkinson, *J. Inorg. Nucl. Chem.*, **3**, 104 (1956).
6. R. B. King and M. B. Bisnette, *Inorg. Chem.*, **6**, 469 (1967).
7. D. Seddon, W. G. Kita, J. Bray, and J. A. McCleverty, *Inorg. Synth.*, **16**, 24 (1976).
8. A. E. Crease and P. Legzdins, *J. Chem. Soc., Dalton*, **1973**, 1501.
9. B. F. G. Johnson and J. A. McCleverty, *Progr. Inorg. Chem.*, 7, 277 (1966).
10. T. S. Piper and G. Wilkinson, *J. Inorg. Nucl. Chem.*, **2**, 38 (1956).
11. R. B. King, *Organometallic Syntheses*, Vol. 1, Academic Press, New York, 1965, pp. 161-163.
12. R. B. King, *Inorg. Chem.*, 7, 90 (1968).
13. M. L. H. Green, T. R. Sanders, and R. N. Whiteley, *Z. Naturforsch.*, **23b**, 106 (1968).
14. R. P. Stewart, Jr., *J. Organometal. Chem.*, **70**, C8 (1974).
15. G. Pass and H. Sutcliffe, *Practical Inorganic Chemistry*, Chapman and Hall Ltd., London, 1968, pp. 145-146, *Inorg. Synth.*, **1**, 55 (1939).

22. RECOVERY OF IRIDIUM FROM LABORATORY RESIDUES

$$\text{Ir compounds} \xrightarrow{\text{heat}} \text{Ir}$$

$$\text{Ir} + 2\text{NaCl} + 2\text{Cl}_2 \xrightarrow{\text{heat}} \text{Na}_2[\text{IrCl}_6]$$

$$\text{Na}_2[\text{IrCl}_6] + 2\text{NH}_4\text{Cl} \longrightarrow (\text{NH}_4)_2[\text{IrCl}_6]\downarrow + 2\text{NaCl}$$

Submitted by GEORGE B. KAUFFMAN* and ROBIN D. MYERS*
Checked by LARRY HALL† and LARRY G. SNEDDON†

Although various procedures for recovering iridium appear in the literature, they are not generally applicable to laboratory residues. Those intended for ores assume the presence of other platinum metals and are therefore unnecessarily

*Department of Chemsitry, California State University, Fresno, CA 93740.
†Department of Chemistry, University of Pennsylvania, Philadelphia, PA 19174.

complicated. Simpler methods involving treatment with reducing agents, on the other hand, reduce other inactive metals, such as copper and silver. Furthermore, standard reduction methods are seldom directly applicable to the more-stable iridium complexes.

The following procedure, modified from the Werner and de Vries' method[1] and Kauffman and Teter's synthesis of ammonium hexachloroiridate(IV),[2] is intended for the recovery of iridium from residues containing base metals and noble metals (other than those of the platinum group), as well as strong complexing agents. The authors have tested the procedure with both actual laboratory residues and synthetic iridium mixtures containing as much as 50% of the following combined impurities: aluminum, ammonium, cobalt, chromium, copper, iron, mercury, potassium, silver, and sodium ions, as well as ethylenediamine, diethyl sulfide, pyridine, tributylphosphine, urea, and thiourea.

Procedure

The dried residue is heated to redness in a large porcelain casserole or evaporating dish over a Meker burner, and ignition is continued until fuming has ceased. This process removes ammonium salts and volatile substances and also decomposes iridium complexes to metallic iridium. (■ **Caution.** *All ignitions and evaporations should be carried out under the hood with adequate shielding!)* If perchlorate or nitrate is present along with organic materials in the residue, an explosion may occur on heating. A test ignition of a gram or so of residue should be carried out before larger-scale ignitions are attempted.

The ignited residue is treated with two times its volume of concentrated hydrochloric acid and evaporated to dryness and absence of fumes several times. Most of the soluble salts are removed by thoroughly washing and suction filtering (Büchner funnel) the ignited and powdered residue with five times its volume of boiling water. This washing and filtering operation should be repeated several times. If potassium salts are present in large amounts, insoluble potassium hexachloroiridate(IV) will be formed when the mixture of the residue and sodium chloride is chlorinated. Iridium is not attacked by ordinary laboratory reagents. Use of a sintered-glass funnel to remove insoluble material is not recommended because of difficulty in removing the Ir from the pores. The residue is then washed and suction filtered through the same Büchner funnel with five times its volume of dilute (6 *N*) nitric acid to remove silver and base metals. This washing with acid and filtering operation should be repeated several times.

The residue is then heated to dryness, ignited, and weighed. This residue (*A*) is ground *intimately* with twice its weight of *finely powdered* sodium chloride, and the mixture is spread out in the center of a Pyrex or Vycor combustion tube. (Pyrex may soften slightly at the temperature to be employed.) A porcelain combustion boat fastened to a length of rigid wire is convenient for inserting the sample and dumping it into the tube without loss. The tube is heated in a tube furnace to about 625°. A stream of chlorine is passed slowly through the tube for approximately 2 hours. (*Hood!*) A slight positive gas pressure should be maintained at the exit tube, which should be immersed in a solution of dilute sodium hydroxide to absorb the excess chlorine. Tygon tubing and rubber stoppers are sufficiently chlorine-resistant to be used for connections.

The cooled melt is transferred to a beaker, using about 15 mL of boiling water per gram of residue *A* to rinse the last traces from the tube. The mixture is heated gently and stirred until all soluble material is dissolved, and the hot, dark-brown mixture is transferred quantitatively through filter paper into a beaker with the aid of a minimum volume of boiling water until the washings are colorless. The water-insoluble residue may be mixed with sodium chloride and rechlorinated to obtain additional iridium.

To ensure that the iridium is completely in the tetrapositive state, the filtrate is boiled gently for about 10 minutes with about 5 mL of *aqua regia* ($1HNO_3$/-$4HCl$) for each gram of residue *A*. Thorough removal of *aqua regia* by boiling prevents oxidation of ammonium ion in the next step. For each gram of residue *A* there is added to the hot solution, a solution of 1 g of NH_4Cl dissolved in about 2.5 mL of boiling water. After the beaker has been allowed to cool to room temperature, it is placed in an ice bath until precipitation apppears to be complete (about 30 min). The small black crystals of ammonium hexachloroiridate(IV) are collected by suction filtration, washed successively with 10 mL each of 5 *M* ammonium chloride, 95% ethanol, and diethyl ether, and then air dried. These volumes of wash liquids are for each gram of residue *A*.

In general, the percentage recovery of iridium increases slightly with increasing percentage of iridium in the residue. For example, with a mixture containing 50% iridium, 86.6% recovery was attained, while with a mixture containing 75% iridium, the recovery rose to 88.7%.* *Anal.* Calcd. for $(NH_4)_2[IrCl_6]$: Ir, 43.58. Found: Ir, 43.60. (Ignition should be carried out in a stream of hydrogen to decompose any IrO_2 to metallic iridium.) Additional ammonium hexachloroiridate(IV) should be obtained by concentrating the mother liquor and the ammonium chloride solution washings by boiling to one-third of their volume and adding additional ammonium chloride as above. Also, the ammonium hexachloroiridate(IV) may be converted to metallic iridium by ignition in a stream of hydrogen.

References

1. A. Werner and O. de Vries, *Ann. Chem.*, **364**, 126 (1909).
2. G. B. Kauffman and L. A. Teter, *Inorg. Synth.*, 8, 223 (1966).

*Recovery values obtained by the checkers amounted to 75.2% and 77.0%, respectively. Yields reported are based on one chlorination.

Chapter Four

A PHOSPHORUS YLIDE AND SOME OF ITS METAL COMPLEXES

H. SCHMIDBAUR*

Ylides are neutral compounds characterized by internally compensating ionic centers, a carbanionic group and a neighboring onium unit, typically localized at phosphorus, arsenic, or sulfur, Ylidic carbanions are strong nucleophiles and show a high affinity for most metals in their various oxidation states. This can be exemplified by the reactions of a simple phosphorus ylide, like trimethylphosphonium methylide (trimethylmethylenephosphorane), that are now known to lead to organometallic compounds with exceptionally stable carbon-to-metal bonds.

Apart from the donor interaction of this ylide with a metal as a simple monodendate ligand (as in structure *a*), it can also be converted into a bidentate, chelating ligand by a metalation or transylidation process (an excess of ylide acts as a strong base in this procedure). These ligands may act either as a bridging or a chelating donor system (as in sturctures *b* and *c*, respectively). Other chelating systems can be derived from double ylides, as described by formulas *d* and *e*.

*Anorganisch-chemisches Institut, Technische Universität München, Arcisstrasse 21, 8000 München 2, Germany.

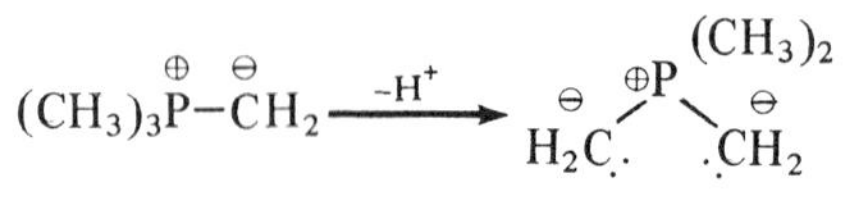

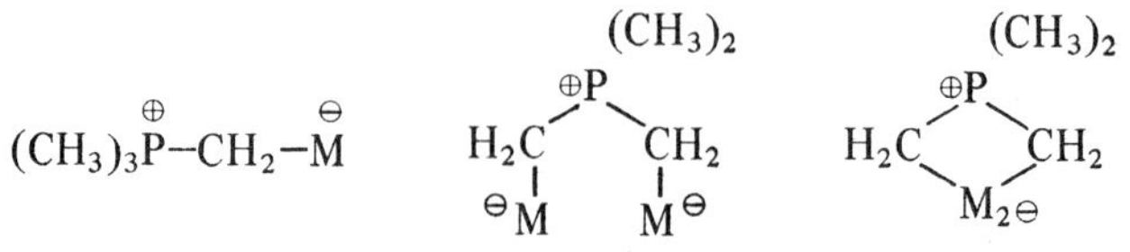

$$(CH_3)_3\overset{\oplus}{P}-\overset{\ominus}{C}H_2 \xrightarrow{-H^+} \overset{\ominus}{H_2C}{:}\!-\!\overset{\oplus}{P}(CH_3)_2\!-\!{:}\overset{\ominus}{C}H_2$$

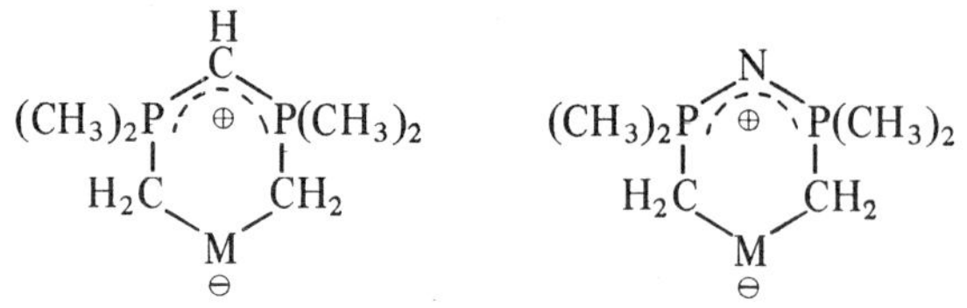

$$(CH_3)_3\overset{\oplus}{P}-CH_2-\overset{\ominus}{M}$$

a

$$(CH_3)_2\overset{\oplus}{P}(CH_2-\overset{\ominus}{M})_2$$

b

$$(CH_3)_2\overset{\oplus}{P}(CH_2)_2M_2^{\ominus}$$

c

$$\left[(CH_3)_2P\overset{\oplus}{\cdots}\overset{H}{C}\cdots P(CH_3)_2\right](CH_2)_2\overset{\ominus}{M}$$

d

$$\left[(CH_3)_2P\overset{\oplus}{\cdots}N\cdots P(CH_3)_2\right](CH_2)_2\overset{\ominus}{M}$$

e

Main-group as well as transition metal acceptor centers of various groups of the periodic table can be incorporated into the resulting novel coordination compounds, where the ylides may function as the sole ligands or a part of a diversified coordination sphere. Coordination numbers so far range from 2 to 8.

The materials described below are some representative examples for the various modes of interaction and show some characteristic properties of this class of compounds. Only the coordination number 2 is covered here, but the reader can easily locate references on other examples in the recent literature, which is compiled at the end of Section 23.

23. TRIMETHYLPHOSPHONIUM METHYLIDE (Trimethylmethylenephosphorane)

Basically, two methods of synthesis are available for $(CH_3)_3PCH_2$, one of which is a three-step process finished by a clean, high-yield desilylation procedure.[1] The second route, proposed more recently,[2] has only two steps, the second of which requires less time but somewhat more effort in the purification of the final product. This reaction is not free of by-products[3], and it is essential to adhere strictly to the conditions given. Both methods are described, because each has certain advantages in the synthesis of homologues.

A. FROM TRIMETHYLPHOSPHONIUM (TRIMETHYLSILYL)METHYLIDE (Trimethyl[(trimethylsilyl)methylene]phosphorane)

$$(CH_3)_3P + (CH_3)_3Si(CH_2Cl) \longrightarrow [(CH_3)_3PCH_2Si(CH_3)_3]Cl \qquad (a)$$

$$[(CH_3)_3PCH_2Si(CH_3)_3]Cl + C_4H_9Li \longrightarrow$$
$$(CH_3)_3P{=}CHSi(CH_3)_3 + C_4H_{10} + LiCl \qquad (b)$$

$$(CH_3)_3P{=}CHSi(CH_3)_3 + CH_3OH \longrightarrow (CH_3)_3P{=}CH_2 + (CH_3)_3Si(OCH_3) \qquad (c)$$

Submitted by H. SCHMIDBAUR* and W. TRONICH†
Checked by N. E. MILLER‡

Procedure

a. In a dry box filled with oxygen-free, dry nitrogen, 38.0 g of trimethylphosphine[4] is added to 61.8 g of (chloromethyl)trimethylsilane[5] (0.5 mole each) in a 250-mL, round-bottomed flask, which is then closed by a stopper and a steel spring and after 2 hours is heated to 30° in a water bath for 8 days. (■ **Caution.** *Trimethylphsophine is toxic and flammable. It should be handled using standard techniques for toxic and air-sensitive compounds.*§) After this period the flask is opened in the dry box and attached to a standard vacuum system. Through pumping (to 1 torr) at 20° the remaining volatiles are removed slowly and collected in a trap cooled with liquid nitrogen. The yield of colorless crystalline material is 82 g (83%).[6] The product is slightly hygroscopic and must be handled in a dry box or glove bag unless the humidity is very low.

b. The phosphonium salt from (a) (50 g, 0.25 mole) is dispersed in 30 mL of diethyl ether with rapid stirring and a solution of an equivalent of *n*-butyllithium in *n*-pentane is added slowly through a dropping funnel. The 250-mL, three-necked flask employed is fitted with a reflux condenser closed by a bubbler filled with oil and flushed with purified nitrogen. *n*-Butane is evolved and the precipitate changes into LiCl. After addition of the organometallic reagent, stirring is continued for 1 hour at 20°, followed by filtration through a glass frit and washing of the LiCl residue with two 15-mL portions of ether. Fractional dis-

*Anorganisch-chemisches Laboratorium, Technische Universität München D 8000 München, Germany.
†Hoechst AG, Frankfurt-Hoechst, Germany.
‡Department of Chemistry, University of South Dakota, Vermillion SD 57069.
§D. F. Shriver, *The Manipulation of Air-sensitive Compounds*, McGraw Hill Book Co., New York, 1969.

tillation of the filtrate yields 38 g of trimethyl[(trimethylsilyl)methylene] phosphorane[6] (92%), bp 66°/12 torr. The checker obtained an 84% yield.

c. In a 250-mL, three-necked flask flushed with nitrogen, 24.6 g of $(CH_3)_3P{=}CHSi(CH_3)_3$ (0.15 mole) is added to 150 mL of diethyl ether and cooled to 0° in an ice bath. With stirring, 4.8 g of methanol dissolved in 150 mL diethyl ether is dropped slowly into this solution over a period of 1-2 hours. Stirring is continued for 3 hours at 20°, and the mixture is subjected to fractional distillation. The minimum yield is 12.6 g (93%), bp 118-120°/750 torr, mp 13-14°. *Anal.* Calcd. for $C_4H_{11}P$ (MW, 90.1): C, 53.32; H, 12.30; P, 34.38: Found: C, 53.12; H, 12.32, P, 33.60. Molecular weight: 91 (cryoscopic in benzene); infrared/Raman analysis: reference 7; ^{1}H, ^{13}C, ^{31}P NMR analysis: reference 1, 8, 9; photoelectron spectroscopy: reference 10; electron diffraction study: reference 11.

B. FROM TETRAMETHYLPHOSPHONIUM BROMIDE

$$(CH_3)_3P + CH_3Br \longrightarrow [(CH_3)_4P]^{\oplus}Br^{\ominus} \qquad \text{(a)}$$

Submitted by H. F. KLEIN*
Checked by W. C. KASKA and J. C. BALDWIN†

Procedure

The reaction vessel is a 1-liter, three-necked flask equipped with a mechanical stirrer, a Dry Ice/acetone reflux condenser, and a 250-mL dropping funnel with a pressure-equalizing tube. Both openings are fitted with a nitrogen inlet, and the whole system is filled with nitrogen. Dry oxygen-free diethyl ether (600 mL) is added and cooled to −30° by means of an external bath. Methyl bromide (bromomethane) (10 g ampule, 0.105 mole) precooled to −50° is added and the dropping funnel containing a solution of 7.6 g $(CH_3)_3P$ (0.100 mole)[4] in 100 mL of diethyl ether is attached to the flask.

■ **Caution.** *Avoid inhaling CH_3Br or contact with the skin. Trimethylphosphine is toxic and flammable. It should be handled using standard techniques for toxic and air-sensitive compounds* (see footnote of p. 137).

The CH_3Br solution is heated by a water bath at 30° until refluxing starts. The trimethylphosphine solution is added dropwise, with stirring (mechanical stirrer or heavy-duty magnetic stirrer), over a period of 1 hour while the reflux is kept low by adding ice to the bath as necessary. At the end of the addition, stirring is

*Anorganisch-chemisches Laboratorium, Technische Universität München D 8000 München, Germany.

†Department of Chemistry, University of California, Santa Barbara, CA 93106.

interrupted and the mixture is allowed to stand for 3 hours at room temperature. The product is isolated by filtration and dried *in vacuo* at 20° for 2 hours. The yield is 16.0 g (94%).

$$[(CH_3)_4P]Br + NaNH_2 \longrightarrow NH_3 + NaBr + (CH_3)_3P{=}CH_2 \qquad (b)$$

Submitted by R. KÖSTER, D. SIMIĆ and M. A. GRASSBERGER*
Checked by W. C. KASKA and J. C. BALDWIN†

Procedure

■ **Caution.** *The pharmacological properties of many alkylphosphorus compounds are not yet know. They should be handled with care.*

In a 250-mL, two-necked flask flushed with nitrogen, 1.5 g of freshly prepared sodium amide[12](38 mmole)‡ and 6.2 g of tetramethylphosphonium bromide (36 mmole) are suspended in 100 ml of tetrahydrofuran.[13] The mixture is heated under reflux for at least 5 hours. The ammonia evolved is trapped in a scrubbing bottle containing 1 *N* H_2SO_4. Titration of the acid indicates that 36 mmole (100%) of NH_3 are formed in the process. Filtration from the sodium bromide formed (3.2 g) and fractional distillation of the filtrate yields 2.9 g (90%), bp, 122°/760 torr. All isolation and purification procedures are also conducted under nitrogen, using glass apparatus with inert gas inlet cocks.

Properties

The product is a colorless liquid at room temperature that fumes in air and is quickly colored brown if impure or on exposure to oxygen and moisture.[1,2] It is a monomer in solution, showing rapid inter- and intramolecular proton exchange.[8,9] ^1H and ^{13}C NMR indicate a high negative charge at the ylidic carbon. According to infrared/Raman studies the P=C bond order is about 1.6.[7] The first ionization potential is extremely low (6.8 eV).[10] The gas-phase structure shows a short P=C bond (1.60 Å).[11]

References

1. H. Schmidbaur and W. Tronich, *Chem. Ber.*, **101**, 595 (1968).
2. R. Köster, D. Simić, and M. A. Grassberger, *Ann. Chem.*, 739, 211 (1970).

*Max-Planck Institut für Kohlenforschung Mülheim/Ruhr, Germany.
†Department of Chemistry, University of California, Santa Barbara, CA 93106.

‡The checker recommends potassium hydride (weighed in a glove bag) to be used instead of $NaNH_2$. The reaction occurs in 1 hour without heating.

3. H. Schmidbaur and H. J. Füller, *Angew. Chem.*, 88, 541 (1976).
4. W. Wolfsberger and H. Schmidbaur, *Synth. Inorg. Metalorg. Chem.*, **4**, 149 (1974), and references therein; *Inorg. Synth.*, **11**, 128 (1968).
5. Commercially available from PCR, Gainesville, FL 32602, and others.
6. This intermediate was fully characterized by two groups: (a) N. E. Miller, *J. Am. Chem. Soc.*, 87, 390 (1965); *Inorg. Chem.*, **4**, 1458 (1965). (b) H. Schmidbaur and W. Tronich, *Chem. Ber.*, **100**, 1032 (1967).
7. W. Sawodny, *Z. Anorg. Allgem. Chem.*, **368**, 284 (1969).
8. H. Schmidbaur, W. Buchner, and D. Scheutzow, *Chem. Ber.*, **106**, 1251 (1973).
9. K. Hildenbrand and H. Dreeskamp, *Z. Naturforsch.*, **28b**, 226 (1973).
10. K. A. Ostoja-Starzewski, H. tom Dieck, and H. Bock, *J. Organometal. Chem.*, **65**, 311 (1974).
11. E. A. V. Ebsworth, D. Rankin, O. Gasser, and H. Schmidbaur, to be published.
12. K. W. Greenlee and A. L. Henne, *Inorg. Synth.*, **2**, 128 (1946).
13. *Inorg. Synth.*, **12**, 317 (1970).

24. YLIDE COMPLEXES OF SOME IB AND IIB METALS

A. BIS[TRIMETHYL(METHYLENE)PHOSPHORANE]MERCURY DICHLORIDE[1] (Bis(trimethylphosphonium methylide)mercury Dichloride)

$$HgCl_2 + 2\,(CH_3)_3P{=}CH_2 \longrightarrow [Hg[CH_2P(CH_3)_3]_2]Cl_2$$

Submitted by H. SCHMIDBAUR and K. H. RÄTHLEIN*
Checked by W. C. KASKA† and J. C. BALDWIN†

Procedure

■ **Caution.** *$(CH_3)_3PCH_2$ is toxic and flammable. It should be handled using standard techniques for toxic and air-sensitive compounds* (see footnote, p. 137).

Under nitrogen 0.153 g of $(CH_3)_3PCH_2$ (0.170 mmole, 0.15 mL) is transferred to 10 mL of diethyl ether using a precision syringe, and this solution is added to a suspension of 0.230 g of mercury(II) chloride (0.850 mmole) in 25 mL of diethyl ether at 0° with stirring by a magnetic bar. A colorless precipitate is formed immediately. Stirring is continued for 3 days in the closed vessel and the product is filtered through a glass frit, washed with two portions of 5 mL of

*Anorganisch-chemisches Laboratorium, Technische Universität München, D 8000 München, Germany.
†Department of Chemistry, University of California, Santa Barbara, CA 93106.

diethyl ether, and dried *in vacuo*. A *minimum* yield of 0.31 g (81%) is obtained. *Anal.* Calcd. for $C_8H_{22}P_2HgCl_2$ (MW 451.71): C, 21.27; H, 4.91, Hg, 42.5. Found: C, 21.22; H, 5.09; Hg, 41.3.

Properties

The compound is a colorless polycrystalline material, decomposition temperature 200°, that is insoluble in all nonpolar organic solvents. It is, however, quickly soluble in water and alcohols, followed by solvolysis. In acidic medium, mercury halides and phosphonium halides are formed in quantitative yields.[1]

An isoelectronic cation, $[Au[CH_2P(CH_3)_3]_2]^{\oplus}$, has also been reported and was shown to exhibit similar properties.[2] Zinc and cadmium halides form polymeric materials with the ylide.[3,4]

B. METHYL[TRIMETHYL(METHYLENE)PHOSPHORANE]GOLD[5] (Methyl(trimethylphosphonium methylide)gold)

$$CH_3Au[P(CH_3)_3] + (CH_3)_3P{=}CH_2 \xrightarrow{-(CH_3)_3P} CH_3Au[CH_2P(CH_3)_3]$$

Submitted by H. SCHMIDBAUR* and R. FRANKE†
Checked by N. E. MILLER‡

Procedure

■ **Caution.** *$(CH_3)_3PCH_2$ is toxic and flammable. It should be handled using standard techniques for toxic and air-sensitive compounds* (see footnote, p. 137).

In a small rubber-capped dropping funnel flushed with nitrogen, 90 mg of $(CH_3)_3P{=}CH_2$ (1.0 mmole, 0.1 mL) is dissolved in 5 mL of diethyl ether by transfer with a precision syringe. At −60° this solution is added to 15 mL of ether containing 288 mg of methyl(trimethylphosphine)gold (1.0 mmole), prepared from $[(CH_3)_3P]AuCl$ and CH_3Li, as described in the literature.[6] A colorless precipitate is formed immediately. The mixture is allowed to warm to 20° with stirring under nitrogen and is filtered through a glass frit. The product

*Anorganisch-chemisches Laboratorium, Technische Universität München D 8000 München, Germany.
†Hoechst AG, D 6230 Frankfurt-Hoechst, Germany.
‡Department of Chemistry, University of South Dakota, Vermillion, SD 57069.

is washed with 5 mL of *n*-pentane and dried in a vacuum. A yield of 240 mg (80%) is obtained. *Anal.* Calcd. for $C_5H_{14}AuP$ (MW., 302.1): C, 19.88; H, 4.67. Found. C, 19.77; H, 4.92. Molecular weight *m/e* 302 (mass spectroscopy).

Properties

The compound $CH_3Au[CH_2P(CH_3)_3]$ is a colorless crystalline solid, mp 119-121° without decomposition (checker observed mp 123-124°, with darkening) and is soluble in benzene and toluene. According to its mass spectrum it is a monomer. Its proton NMR spectrum shows a doublet of triplets for the CH_3Au group through coupling with the CH_2P moiety of the new ligand: $J_{(HCAuCH)}$ = 0.6 Hz, $J_{(HCAuCP)}$ = 1.4 Hz. The ^{31}P chemical shift is 23 ppm (rel. H_3PO_4), a 21-ppm deshielding compared with the uncomplexed ylide. An isoelectronic $[CH_3Hg[CH_2P(CH_3)]]^{\oplus}$ cation is also known.[1] In the infrared spectrum the AuC stretching absorptions are found at 597, 538, and 518 cm^{-1}.

C. BIS-μ-[[DIMETHYL(METHYLENE)PHOSPHORANYL]METHYL]-DISILVER[7]

$$2(CH_3)_3PAgCl + 4(CH_3)_3P{=}CH_2 \longrightarrow 2(CH_3)_3P + 2[(CH_3)_4P]Cl +$$

$$\begin{array}{ccccc} CH_3 & & CH_2{-}Ag{-}CH_2 & & CH_3 \\ & P & & P & \\ CH_3 & & CH_2{-}Ag{-}CH_2 & & CH_3 \end{array}$$

Submitted by H. SCHMIDBAUR* and J. ADLKOFER†
Checked by N. E. MILLER‡

Procedure

Chloro(trimethylphosphine)silver[8] (0.91 g, 1.03 mmole) is suspended in 25 mL of toluene with magnetic stirring and then cooled to −20°. To this is added $(CH_3)_3P{=}CH_2$ (0.74 g, 8.24 mmole, 0.83 mL) in one portion under nitrogen using a micropipette. The mixture is allowed to warm to 20° and stirring is continued overnight under nitrogen. The colorless precipitate is filtered through a glass frit and washed with 5 mL of diethyl ether. It consists largely of the phosphonium salt by-product. The filtrate is concentrated *in vacuo* and cooled

*Anorganisch-chemische Laboratorium, Technische Universität München D 8000 München, Germany.
†BASF AG, D 6700 Ludwigshafen, Germany.
‡Department of Chemistry, University of South Dakota, Vermillion, SD 57069.

to −30°. Colorless crystals are obtained in a yield of 0.75 g (92%). All reaction vessels should be protected against direct irradiation and the nitrogen atmosphere maintained throughout the workup procedure, in either a dry box or a glove bag, or through nitrogen inlet cocks at all glass parts. The checker obtained a slightly yellowish product (67% yield). It was sublimed at 110-140°/high vacuum to give a white material. *Anal.* Calcd for $C_8H_{20}Ag_2P_2$: C, 24.39; H, 5.12. Found: C, 25.02; H, 5.35. Molecular mass: *m/e* 394 (mass spectroscopy).

Properties

The compound bis-μ-[[dimethyl(methylene)phosphoranyl] methyl] disilver, mp 153-155°, is volatile at 150°/0.1 torr, its mass spectrum showing the parent ion. It is soluble in benzene and toluene. The proton NMR spectrum shows a doublet of doublets for the bridging CH_2 groups. This splitting is due to strong *HCP* and *HCAg* couplings, the latter proving the covalent bonding of the ligand to the metal. An analogous coupling (*PCAg*) is observed in the ^{31}P NMR spectrum, which is a triplet in the $\{^1H\}$-experiment. The infrared spectrum shows Ag-C stretching absorptions at 472 and 522 cm^{-1}. The crystal structures of the *copper*[9] and *gold*[10] analogues have been determined. Centrosymmetrical eight-membered ring structures with linear C−M−C groups parallel to each other (M = Cu, Au) have thus been confirmed.

References

1. H. Schmidbaur and K.-H. Räthlein, *Chem. Ber.*, **107**, 102 (1974).
2. H. Schmidbaur and R. Franke, *Angew. Chem.*, 85, 449 (1973), *Angew. Chem. Int. Ed. (Engl.)*, **12**, 416 (1973), *Chem. Ber.* **108** 1321 (1975).
3. H. Schmidbaur, *Acc. Chem. Res.*, **8**, 62 (1975).
4. H. Schmidbaur and J. Eberlein, *Z. Anorg. Allgem. Chem.* in preparation.
5. H. Schmidbaur and R. Franke, *Chem. Ber.*, **108**, 1321 (1975), Dissertation R. Franke, Univ. Würzburg 1974.
6. H. Schmidbaur and A. Shiotani, *Chem. Ber.*, **104**, 2821 (1971) and references therein.
7. H. Schmidbaur, J. Adlkofer, and W. Buchner, *Angew. Chem.*, 85, 448 (1973). H. Schmidbaur, J. Adlkofer, and M. Heimann, *Chem. Ber.*, **107**, 3697 (1974); D. S. Rustad, T. Birchall, and W. L. Jolly, *Inorg. Synth.* **11**, 128 (1968).
8. H. Schmidbaur, J. Adlkofer, and K. Schwirten, *Chem. Ber.*, **105**, 3382 (1972).
9. G. Nardin, L. Randaccio, and F. Zangrando, *J. Organometal. Chem.*, 74, C23 (1974).
10. H. Schmidbaur and R. Franke, *Inorg. Chim. Acta*, **13**, 84 (1975); H. Schmidbaur, J. R. Mandl, A. Frank, and G. Huttner, *Chem. Ber.* **109**, 466 (1976); H. Schmidbaur, J. R. Mandl, W. Richter, V. Bejenke, A. Frank, and G. Huttner, *Chem. Ber.*, **110**, 2236 (1977).

Chapter Five

BORON AND ALUMINUM COMPOUNDS

25. HALODIBORANES(6) AND HALODIBORANES(6)-d_5

Submitted by JOHN E. DRAKE,* BERNARD RAPP,† CHRIS RIDDLE*
and JAMES SIMPSON‡
Checked by TERRANCE MAZANAC§ and SHELDON G. SHORE§

The halodiboranes, B_2H_5Cl, B_2H_5Br, and B_2H_5I, have been prepared by reaction of diborane(6), B_2H_6, with hydrogen halide,[1-3] free halogen,[3-5] and perhaloboranes.[2,6-8] The chloride and bromide are also formed in the hydrogenation of boron trichloride, BCl_3, and boron tribromide BBr_3.[9,10] Monohalodiboranes are of particular interest for physical studies of borane derivatives and in the preparation of adducts.[11] The decomposition of pure monohalodiboranes is relatively slow. Under normal laboratory conditions, however, decomposition proceeds sufficiently rapidly that it is a problem. The chloride and bromide form a series of products (e.g., BH_2Cl, $BHBr_2$) en route to diborane(6) and the trihalide, while the iodide yields diborane(6) and boron triiodide directly.

The most convenient preparation of the bromide and iodide is the reaction between diborane(6) and the corresponding trihaloborane. Hydrogen halide is not present at any stage of the syntheses and this makes trap-to-trap separation of the produ. ιs a simple matter.

*Department of Chemistry, University of Windsor, Windsor, Ontario, Canada N9B 3P4.
†Department of Chemistry, Lewis University, Lockport, IL.
‡Department of Chemistry, University of Otago, Dunedin, New Zealand.
§Department of Chemistry, Ohio State University, Columbus, OH 43210.

An advantage of the procedures described below is that they may be applied to the deuterium derivatives, B_2D_5Br and B_2D_5I, by substituting B_2D_6 for B_2H_6 in the reaction mixtures.

A. BROMODIBORANE(6) AND BROMODIBORANE(6)-d_5

$$5B_2H_6(B_2D_6) + 2BBr_3 \longrightarrow 6B_2H_5Br(B_2D_5Br)$$

■ **Caution.** *The boron hydride derivatives encountered in this preparation are extremely air sensitive and can inflame violently in the atmosphere. They are also very toxic, or believed to be so. The preparation should be carried out in a vacuum line in a well-ventilated area. A grease-free system should be used.* *

Procedure

The reaction vessel consists of a 300-mL glass bulb with cold finger, fitted with a 4-mm-bore greaseless stopcock, through which it may be attached to the vacuum line. Diborane(6)†, B_2H_6, (9.0 mmole) (or diborane(6)-d_6,B_2D_6) is condensed into the evacuated reaction vessel at $-196°$ (liquid nitrogen trap). Freshly distilled‡ boron tribromide, BBr_3 (1.13 g; 4.5 mmole) is added, also at $-196°$, and the reaction vessel is then allowed to warm slowly to $0°$. It is maintained at $0°$ for 3 hours. The reaction vessel is then cooled to $-196°$ and opened to a series of three traps at -78 (Dry Ice/methanol slush), -126 (methylcyclohexane slush) and $-196°$, in that order. Any traces of hydrogen are pumped away through the traps, which are then closed to the pump. The reaction vessel is allowed to warm to room temperature and the products are thereby fractionated. Bromodiborane(6) is retained in the trap at $-126°$. The contents of the traps at -78 and $-196°$ are returned to the reaction vessel and the reaction and separation cycles are repeated. The crude product from the trap at $-126°$ is held in a separate storage vessel at $-196°$. Up to nine successive reaction cycles may be performed before the yield becomes negligible. In this way over 80% of the diborane(6) is converted and recovered as B_2H_5Br (or B_2D_5Br).

*The techniques involved are discussed in detail by D. F. Shriver, in *The Manipulation of Air-sensitive Compounds,* McGraw Hill, Book Co., New York, 1969, pp. 229.

†B_2H_6, may be prepared according to methods given in *Inorg. Synth.*, **11**, *15 (1968)* and **15**, 142 (1974). Also see the methods of K. C. Nainan and G. E. Ryschkewitsch, *Inorg. Nucl. Chem. Lett.*, **6**, 765 (1970) and G. F. Freeguard and L. M. Long, *Chem. Ind. (London)*, **11**, 471 (1965).

‡A middle fraction condensing in a trap at $-78°$ is satisfactory. It should have a vapor pressure of 19.0 torr at $0°$. Boron tribromide may be prepared as in *Inorg. Synth.*, **12**, 146 (1970) or the commercial product (Alfa Inorganics, Ventron Corporation, Danvers, MA 01923) may be used.

The combined fractions of crude B_2H_5Br (or B_2D_5Br) are finally purified by repeated passage through a trap at −96° (toluene slush) into a trap at −196°. Pure B_2H_5Br (or B_2D_5Br) collects in the trap at −196°, while traces of $BHBr_2$ (or $BDBr_2$) remain in the trap at −96°. The reaction may be scaled down conveniently by as much as tenfold. If an increased amount of bromodiborane(6) is desired, the pressure in the reaction vessel may be adjusted at the start of every third cycle by the addition of more reaction mixture, to the initial reaction pressure. This pressure should not exceed 1.5 atm.

■ **Caution.** *Diborane(6) and other borane species in the vacuum line trap may be diluted with nitrogen and vented into a solution of pyridine in benzene for safe disposal. Alternatively, pyridine may be condensed in the vacuum-line trap.*

Properties

Bromodiborane(6) condenses in a vacuum system as a colorless liquid. It has a vapor pressure[3] of 41 torr at −45° and is best characterized by its infrared spectrum.[6] It should be stored at −196° in a grease-free ampule. Before use, its purity should always be checked by vapor pressure determination and the absence of impurity bands in the infrared spectrum. If necessary, refractionation must be performed to remove traces of the usual impurities, which are B_2H_6, $BHBr_2$, and BBr_3. Characteristic infrared bands (cm^{-1}) are as follows:

B_2H_5Br: 2615 (s), 2530 (s), 1565 (vs), 1490 (s), 1163/1150 (s), 1066/1051 (vs), 956/942,907 (m), 818/807 (m).

B_2D_5Br: 1980 (s), 1870 (s), 1175 (vs), 1112/1100 (m), 915 (s), 850/836 (vs), 600 (m), 420 (m).

B. IODODIBORANE(6) AND IODODIBORANE(6)-*d*$_5$

$$5B_2H_6(B_2D_6) + 2BI_3 \longrightarrow 6B_2H_5I(B_2D_5I)$$

■ **Caution.** *See Synthesis A.*

Procedure

The general procedure is given in Section 25-A. The reaction vessel should be covered to exclude light and the mercury manometers closed except when the diborane is measured. These two precautions lessen the chance of extensive product decomposition.

Boron triiodide BI_3, is a solid. The commercial product* may be purified best prior to the reaction by preparing a near-saturated solution of BI_3 in benzene. This should be carried out in a nitrogen-filled glove bag.† Free iodine is then removed by shaking with mercury. The resulting solution is then decanted and benzene distilled out at room temperature on this line. Pure BI_3 (1.57 g; 4.0 mmole) is placed in the reaction vessel. This is done in the glove bag, with a spatula, with the stopcock removed from the reaction vessel.‡ Diborane(6) (10.0 mmole) is condensed into the evacuated vessel at $-196°$. As in Section A, the reaction is allowed to proceed for 3 hours at 0°. The products are then distilled through traps at -45 (chlorobenzene slush) and $-96°$ (toluene slush) into one at $-196°$ (liquid nitrogen). Diborane(6) condenses in the trap at $-196°$ and is returned to the reaction vessel together with the contents of the trap at $-45°$. Most of the unreacted BI_3 remains in the reaction vessel. The trap at $-96°$ contains B_2H_5I (or B_2D_5I), which is transferred to a storage vessel and held at $-196°$. The reaction and separation cycle are repeated until all BI_3 has reacted. During the synthesis, a liquid, which has been shown to be a solution of B_2H_5I in BI_3.[7] is observed in the reaction vessel. By this procedure about 60% of the initial diborane(6) is recovered as B_2H_5I (or B_2D_5I). No other volatile products are observed and final purification of the crude product is achieved in a single distillation through traps at -45, -96, and $-196°$. Pure B_2H_5I (or B_2D_5I) is retained in the trap at $-96°$.

■ **Caution.** *Diborane(6) and other borane species in the vacuum line trap may be diluted with nitrogen and vented into a solution of pyridine in benzene for safe disposal. Alternatively pyridine may be condensed into the vacuum-line trap.*

Properties

Iododiborane(6) condenses in a vacuum system as a colorless liquid. It hydrolyzes rapidly and should be kept out of contact with light and mercury. It should be stored at $-196°$. The vapor pressure[3] of B_2H_5I is 70 torr at 0°. Both B_2H_5I and B_2D_5I are best characterized by their infrared spectra.[7] Characteristic infrared bands (cm^{-1}) are as follows:

B_2H_5I: 2630/2610 (s), 2540 (s), 2528 (s), 1735 (m), 1580 (vvs), 1440 (m), 1165/1151 (s), 1038/1020 (vs), 810/800 (m), 463 (s).

*Alfa Inorganics, Ventron Corporation, Danvers, MA 01923.

†A suitable bag is available from I^2R Co., 108 Franklin Ave., Cheltenham, PA 19012.

‡It is equally satisfactory to pipet the benzene solution of BI_3 directly into the (weighed) reaction vessel. Benzene is pumped off and the amount of BI_3 present is determined by weight difference.

B_2D_5I: 1985 (s), 1863 (s), 1851 (s), 1168 (vs), 899 (m), 828/815 (vs), 388 (m).

References

1. H. W. Myers and R. F. Putnam, *Inorg. Chem.*, **2**, 655 (1963).
2. H. I. Schlesinger and A. B. Burg. *J. Am. Chem. Soc.*, **53**, 4321 (1931).
3. A. Stock and E. Pohland, *Chem. Ber.*, **59**, 2223 (1926).
4. A. Stock, *Chem. Ber.*, **47**, 3115 (1914).
5. A. Stock, E. Kuss, and A. Prees, *Chem. Ber.*, **46**, 1959 (1913).
6. S. B. Rietti and J. Lombardo, *J. Inorg. Nucl. Chem.*, **27**, 247 (1965).
7. J. Cueilleron and J. Mongeot, *Bull. Soc. Chim. Fr.*, **1967**, 1065.
8. J. Cueilleron and J. Reymonet, *Bull. Soc. Chim. Fr.*, **1967**, 1370.
9. H. I. Schlesinger and A. B. Burg, *J. Am. Chem. Soc.*, **53**, 1321 (1931).
10. J. Cueilleron and J. Reymonet, *Bull. Soc. Chim. Fr.*, **1967**, 1367.
11. J. E. Drake and J. Simpson, *Inorg. Chem.*, **6**, 1984 (1967).

26. SODIUM DIHYDRIDOBIS(2-METHOXYETHOXO)-ALUMINATE(1−)

$$Na + Al + 2CH_3OCH_2CH_2OH \xrightarrow[150^\circ C]{H_2,\ 70\text{-}250\ atm}$$

$$Na[AlH_2(OCH_2CH_2OCH_3)_2]$$

Submitted by BOHUSLAV ČÁSENSKÝ,[†] and JIŘÍ MACHÁČEK[†]
Checked by E. C. ASHBY[‡] and H. S. PRASAD[‡]

Methods for the preparation of sodium dihydridobis(2-methoxyethoxo)-aluminate(1−) may be divided into two groups differing in the source of hydrogen. In the methods of the first group the hydride hydrogen is present in the reaction mixture in the form of a hydride.[1] The reactions of trisodium hexahydridoaluminate(3−) with tris(2-methoxyethoxo)aluminum or sodium tetrahydridoaluminate(1−) with (*a*) 2-methoxyethanol, (*b*) 2-methoxyethyl methoxyacetate and (*c*) sodium tetrakis(2-methoxyethoxo)aluminate(1−) may serve as examples. Methods in which the hydride hydrogen is formed by means of the hydrogenation of aluminum[2,3] (direct synthesis) belong to the second group. In

*Institute of Inorganic Chemistry, Czechoslovak Academy of Sciences, 250 68 Rez near Prague, Czechoslovakia.

†School of Chemistry, Georgia Institute of Technology, Atlanta, GA 30332.

this case the synthesis from sodium, aluminum, and hydrogen, and (*a*) 2-methoxyethanol, (*b*) sodium tetrakis(2-methoxyethoxo)aluminate(1−), and (*c*) tris(2-methoxyethoxo)aluminum may be mentioned.

Although all methods give nearly theoretical yields when work is carried out in an inert and water-free medium, the direct synthesis from sodium, aluminum, and 2-methoxyethanol under hydrogen pressure is most suitable because of its simplicity and easy accessibility of the starting materials. The synthesis is accomplished in a high-pressure hydrogenation vessel tested to the minimum pressure of 250 atm at 200° and equipped with a heater, manometer, pyrometric tube, and valve. The method of stirring the reaction mixture is arbitrary, but aluminium particles must be held in suspension. A rotating autoclave, an agitating autoclave, or an autoclave equipped with a stirrer may be used.

Sodium, aluminum powder, 2-methoxyethanol, and hydrogen serve as the reactants, benzene or toluene as the solvent, and nitrogen as the inert gas.

The most advantageous reaction conditions for the preparation are as follows: a reaction temperature of 150° ± 5°, minimum hydrogen pressure of 70 atm, a final concentration of the product of 50-55%, and a 50% theoretical excess of the finest aluminum powder.

■ **Caution.** *As hydrogen is liberated before hydrogenation starts when 2-methoxyethanol reacts with sodium and aluminium, the volume of the reaction mixture must not exceed 55% of the total free volume of the autoclave and the maximum pressure of hydrogen must not exceed the permitted working pressure. This part of the preparation is exothermic and the temperature of the reaction mixture must not exceed 170°. When more than 15 moles of sodium dihydridobis(2-methoxyethoxo)aluminate(1−), is prepared it is necessary to control this reaction by slow addition of 2-methoxyethanol to sodium and aluminum in the closed autoclave.*

Procedure

Aluminum (28.1 g, 1.02 moles of 98% Al) in the form of a powder with surface area of 0.180 m^2/g, anhydrous benzene (128.5 g), and anhydrous 2-methoxyethanol (104.5 g, 1.375 moles) are placed in a 0.5-L stirred autoclave (paddle-wheel stirrer) and sodium (16.1 g, 0.7 mole) is added slowly in small pieces. The autoclave is closed and pressurized to 85 atm. The reaction mixture is warmed to 40°, and the pressure reaches 130 atm. The exothermic reaction with evolution of hydrogen starts at 115-120°. The measured pressure is converted according to Charles law to 0°. Within 20 minutes the temperature rises to 140° and the pressure to 218 atm (114 atm at 0°). Until this exothermic reaction, indicated by a pressure increase, is finished, the autoclave must not be heated, because the product could be damaged by temperatures above 170°. Pressure and temperature are read at 3-minute intervals. When the maximum value of the pressure (converted to 0°) is reached, hydrogenation starts. At that moment the consumption of hydrogen is also calculated. Pressure and temperature are read at 15-20 minute intervals. The temperature is held constant at 150° and the reaction mixture is stirred as long as hydrogen is consumed (4-6 hr). The total consumption of hydrogen is 69 atm (converted to 0°). This value is for the specified quantities of starting materials and volume of the autoclave.

After the autoclave is cooled and vented, the stirred reaction mixture (solution of the product in benzene, aluminium in excess, impurities) is sucked by a polyethylene tube into a 0.5-L flask. This operation and those following are carried out in a protective nitrogen atmosphere. The solution of the product in benzene is separated from the reaction mixture by filtration (or decantation). A medium-porosity fritted-glass filter is used for the filtration (see Figure 3).

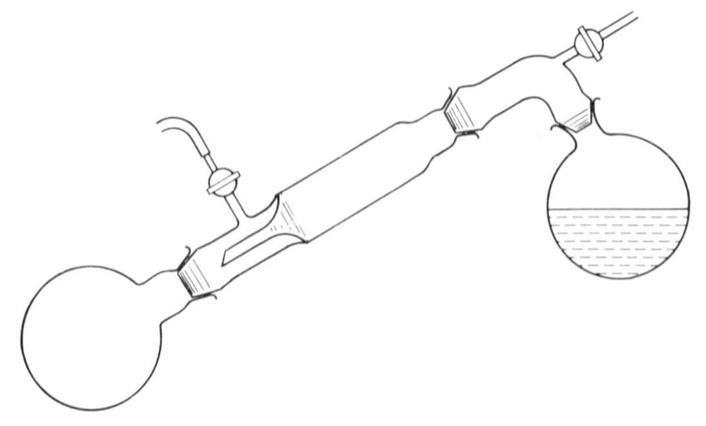

Fig. 3.

Considerably improved filtration may be accomplished when the fritted-glass filter is covered with a 1-1.5 cm layer of powdered aluminum (or dried Celite filter aid). Before starting the filtration, the whole apparatus is evacuated. Then the reaction mixture is poured into the fritted-glass filter while nitrogen is introduced slowly into the space above the filter. The solids are washed with 15 ml of benzene and the filtrates are combined to give 262 g of clear yellowish solution. The yield with respect to the initial amount of 2-methoxyethanol is 98.1%. *Anal.* Calcd. for $Na[AlH_2(OCH_2CH_2OCH_3)_2]$: Na, 11.37; Al, 13.35; H , 0.997%. Found: Na, 6.04; Al, 6.94; H^-, 0.537%, Na:Al:H^- = 1.02:1:2.07.

Properties

Sodium dihydridobis(2-methoxyethoxo)aluminate(1−) is a clear or light-yellow viscous liquid (20°) with d_{20} = 1.122 g/cm^3. Upon cooling to 0-5°, it solidifies to a vitreous brittle substance without a sharp melting point.[1] This compound is exceptionally soluble[1,4] in aromatic hydrocarbons (unlimited solubility in benzene). At 0° in toluene, xylene, and mesitylene it forms two conjugate solutions with the concentrations 5.73 and 41.72, 5.51 and 62.04, and 2.88 and 63.7%, respectively. With diethyl ether it forms two conjugate solutions at 0° with the concentrations of 9.41 and 67.51%. Unlimited solubility was observed with 1,2-dimethoxyethane and tetrahydrofuran. It is insoluble in aliphatic (heptane) and alicyclic (cyclohexane) hydrocarbons.

The analytical determination of aluminium (complexometrically after the hydrolysis) and of hydride hydrogen (volumetrically after decomposition with an acid) is sufficient for the determination of the concentration and the quality of the product in the solution.[5]

Sodium dihydridobis(2-methoxyethoxo)aluminate(1−) is a derivative of $Na[AlH_4]$ and its behavior is chemically similar to that of the tetrahydridoaluminates, $Na[AlH_4]$ and $Li[AlH_4]$.[6] The compound is stable up to 170°, when decomposition begins; the decomposition is spontaneous at 214°.[1]

A valuable property of sodium dihydridobis(2-methoxyethoxo)aluminate(1−), in addition to its solubility in aromatic hydrocarbons, is the nonpyrophoricity both in air and in contact with water. Danger of self-ignition arises only when concentrated solutions of this compound (70% and more) are brought into contact with dry textiles. A spilled solution may be wiped off without any danger with a wet cloth.

References

1. B. Čásenský, J. Macháček, and K. Abrham, *Collect. Czech. Chem. Commun.*, **36**, 2648 (1971).
2. B. Čásenský, J. Macháček, and K. Abrham, *Collect. Czech. Chem. Commun.*, **37**, 1178 (1972).
3. B. Čásenský, J. Mackáček, and K. Abrham, *Collect. Czech. Chem. Commun.*, **37**, 2537 (1972).
4. J. Duben, B. Čásenský, and O. Štrouf, *Collect. Czech. Chem. Commun.*, **39**, 546 (1974).
5. H. Plotová, M. Skalický, and M. Filip, *Chem. Průmysl*, **23**/48, 116 (1973).
6. J. Málek, and M. Černý, *Synthesis*, **1972** (5), 217.

Chapter Six

GERMANIUM HYDRIDE DERIVATIVES

27. BROMOTRIMETHYLGERMANE

$$(CH_3)_4Ge + Br_2 \xrightarrow[\text{1-bromopropane}]{} (CH_3)_3GeBr + CH_3Br$$

Submitted by WITTA PRIESTER* and ROBERT WEST*
Checked by HOWARD ZINGLER† and CHARLES H. VAN DYKE†

There is a variety of reported routes to the halotrimethylgermanes.[1-4] These procedures have caused considerable difficulty in that they are often irreproducible, require long reaction times, involve sealed-tube reactions that are difficult to perform on a large scale, or produce only fair yields of products. The following procedure, a modification of one used by Mironov and Kravchenko,[5] circumvents these problems.

Procedure

■ **Caution.** *All operations should be carried out in a well-ventilated area because of the toxicity of bromine and the germane derivatives.*

Bromine (35 g, 11.3 mL, 0.22 mole) is added dropwise over a period of 1 hour to 26.4 g (0.20 mole) of tetramethylgermane[6] mixed with 15 mL of 1-bromopropane in a flame-dried 250-mL, three-necked flask, equipped with a magnetic stirrer,

*Department of Chemistry, University of Wisconsin, Madison, WI 53706.
†Department of Chemistry, Carnegie-Mellon University, Pittsburgh, PA 15213.

a stoppered pressure-equalizing dropping funnel, thermometer, and water-cooled reflux condenser with $CaCl_2$ drying tube at the top. The resulting mixture is heated to reflux for 16 hours, during which time the pot temperature rises to 80-90°. After cooling to room temperature, approximately 5 mL of mercury is added, with stirring, to remove the excess bromine. The clear solution is decanted into a 100-mL flask. Fractional distillation through a Vigreux column yields 25-31 g (65-80%) of pure bromotrimethylgermane, bp 112-113°.

Properties

Bromotrimethylgermane is a water-sensitive, clear liquid, the density of which is 1.544 g/mL at 18°.[1] It is generally soluble in organic liquids and its ^1H NMR spectrum in CCl_4 has a singlet at δ0.84.

References

1. L. M. Dennis and W. I. Patnode, *J. Am. Chem. Soc.*, **52**, 2779 (1930).
2. H. Sakurai, K. Tominaga, T. Watanabe, and M. Kumada, *Tetrahedron Lett.*, 5493 (1966).
3. R. E. J. Bichler, M. R. Booth, H. C. Clark, and B. K. Hunter, *Inorg. Synth.*, **12**, 64 (1970).
4. E. W. Randall and J. J. Zuckerman, *J. Am. Chem. Soc.*, **90**, 3167 (1968).
5. V. F. Mironov and A. C. Kravchenko, *Izv. Akad. Nauk S.S.S.R., Ser. Khim.*, **1965**, 988.
6. Tetramethylgermane was prepared by the method of Brooks and Glockling, *Inorg. Synth.*, **12**, 58 (1970); it is also available from Alfa Products, Ventron Corporation, Danvers, MA, 01923.

28. DIMETHYLGERMANE AND MONOHALODIMETHYLGERMANES

Submitted by J. E. DRAKE,* B. M. GLAVINCEVSKI,* R. T. HEMMINGS,† and H. E. HENDERSON*
Checked by J. D. ODOM‡ and E. J. STAMPF‡

The germanium-halogen bond is particularly labile and has played an important role in germanium chemistry. Synthetic routes to the fully substituted halo-

*University of Windsor, Windsor, Ontario, Canada N9B 3P4.
†University of the West Indies, Mona, St. Andrew, Jamaica, W. I.
‡Univeristy of South Carolina, Columbia, SC 29208.

organogermanes are generally well defined,[1] but preparations of the related hydrido species have appeared only comparatively recently. Sufficient impetus for the development of high-yield syntheses arises from the rather high cost of germanium required for these organometallics. The preparations of dimethylgermane and its monohalo derivatives[2] are illustrative of the techniques involved. The routes are suitable for hydrido-alkylgermanes in general.[3]

In the past dimethylgermane has been prepared by the sodium tetrahydroborate(1−) reduction of $(CH_3)_2GeBr_2$ in aqueous acidic solution[4]; chlorodimethylgermane (*I*) by the $AlCl_3$ catalyzed reaction of gaseous dimethylgermane with hydrogen chloride[5]; bromodimethylgermane (*II*) by halogen exchange of (*I*) with hydrogen bromide[6]; fluorodimethylgermane (*III*) from (*II*) with lead(II) fluoride;[6] and iododimethylgermane (*IV*) from 1,1,3,3-tetramethyldigermoxane ($[(CH_3)_2GeH]_2O$), with hydrogen iodide.[7] We describe alternative and more convenient routes to (*I*) and (*IV*) by halogenation of dimethylgermane with boron trichloride and iodine, respectively, and by conventional halide metathesis of (*I*) with hydrogen bromide and (*IV*) with lead(II) fluoride. The last reaction proceeds more readily than that with the bromide.[6] We also describe the lithium tetrahydridoaluminate(1−) reduction of $(CH_3)_2GeCl_2$ in nonaqueous solution, which we have found applicable to the syntheses of organosilanes and germanes in general where sodium tetrahydroborate(1−) in aqueous media cannot be tolerated. Each preparation may be accomplished in approximately 2 hours.

Starting Materials

All reactions are carried out using conventional vacuum-line techniques.[8] Teflon-in-glass stopcocks and a silicone lubricant for ground-glass joints should be used because of the marked solubility of germanes in hydrocarbon greases.

Dichlorodimethylgermane (Laramie Chem. Co., WY) is degassed at −45° (chlorobenzene slush); its purity may be confirmed by ^1H NMR[2] or vibrational[9] spectroscopy. Hydrogen bromide (Matheson Gas Products, East Rutherford NJ 07073) is passed through traps at −78° (acetone-solid CO_2) and degassed at −196°; its purity is checked conveniently by its vapor pressure at −78°/400 torr.[10] Boron trichloride (Matheson) is degassed at −112° (1-bromobutane slush) and the purity of middle fractions is confirmed by its infrared spectrum[11] and its vapor pressure at −45°/50 torr.[12] Lead(II) fluoride (Allied Chemical, Morristown, NJ) is dried at 50° in a high vacuum for several days before use, or it may be prepared.[13]

■ **Caution.** *The germanium compounds encountered in these preparations are of unknown toxicity. Manipulations should be carried out in a vacuum system in a well-ventilated area.*

A. DIMETHYLGERMANE

$$2(CH_3)_2GeCl_2 + LiAlH_4 \longrightarrow 2(CH_3)_2GeH_2 + LiCl + AlCl_3$$

■ **Caution.** *See the recommendations in the general caution above.*

Procedure

Fresh lithium tetrahydridoaluminate(1−), $LiAlH_4$ (about 2 g), is placed in a N_2-purged tipping tube fitted to one joint of a 250-mL, two-necked, round-bottomed flask containing a magnetic stirring bar. The other joint is connected to the vacuum line by means of a conventional low-temperature reflux condenser.*

The system is evacuated thoroughly. Anhydrous dibutyl ether (about 25 mL) and $(CH_3)_2GeCl_2$ (1.9431 g, 11.14 mmole) are distilled into the flask. The condenser is maintained at −78°. The mixture is stirred and cooled with an ice bath while the $LiAlH_4$ is added SLOWLY from the tipping tube (about 0.2-g portions). A vigorous effervescence should be apparent immediately, and crude $(CH_3)_2GeH_2$ is pumped from the reaction vessel as it is formed into a series of four traps held at −196°. When all the $LiAlH_4$ has been added, the ice bath is removed and pumping is continued until there is no further sign of gas evolution. This usually takes no more than 5 minutes if fresh $LiAlH_4$ has been used. The crude $(CH_3)_2GeH_2$ is passed through traps at −78° (two), −126° (methylcyclohexane slush), and −196°. The contents of the trap at −126° are refractionated through a trap at −95° (toluene slush) when pure $(CH_3)_2GeH_2$ (10.3 mmole) is obtained in a trap at −196° which follows.† Yields based on $(CH_3)_2GeCl_2$ are typically 95%.

■ **Caution.** *The residue remaining in the reactor should be treated cautiously with 2-propanol before disposal in an efficient fume hood.*

Properties

Dimethylgermane is a colorless, highly volatile liquid (mp −144°, bp 3.0°). The vapor pressures at −95, −45. and −23° (CCl_4 slush) are 1.7, 80.6, and 256 torr, respectively.[4] The proton chemical shifts (downfield from TMS, 5% v/v in CCl_4) are δCH_3 (triplet) 0.29 ppm and δGeH (septet) 3.73 ppm with J^{vic}_{HH} 3.95 Hz.[14]

*See reference 8 p. 124 for a similar apparatus.

†The checkers report the formation of a very small amount of CH_3GeH_3 that was difficult to remove from $(CH_3)_2GeH_2$ by trap-to-trap fractionation. The $(CH_3)GeH_3$ is characterized by its ^1H NMR spectrum,[14] the typical A-type contour at 1258 cm^{-1} in the infrared spectrum,[15] and $(CH_3)_2GeH_2$ vapor pressures[4] that are higher than expected. The CH_3GeH_3 may have arisen from the reduction of CH_3GeCl_3, which is a likely trace impurity in some commercial grades of $(CH_3)_2GeCl_2$.

Sharp bands are observed at 2987, 2936, 2080 (vs), 2062 (vs), 889, 843 (vs), 604 (vs) cm^{-1}. in the gas-phase infrared spectrum.[15] Dimethylgermane may be stored at room temperature in the gas phase.

B. CHLORODIMETHYLGERMANE

$$6(CH_3)_2GeH_2 + 2BCl_3 \longrightarrow 6(CH_3)_2GeHCl + B_2H_6$$

■ **Caution.** *B_2H_6 may explode on contact with air. It should be distilled into a large excess of $(C_2H_5)_3N$ before removal from the vacuum line. See the recommendations in the general cautions on pp. 146 and 147.*

Procedure

The reactor is a 200-mL flask attached to the vacuum system by a ground-glass joint. After thorough evacuation BCl_3 (3.25 mmole) and $(CH_3)_2GeH_2$ (about 10.5 mmoles) are distilled into the vessel held at −196°. The reactor is isolated from the vacuum system (a Teflon-in-glass stopcock attached to the reactor is convenient). The mixture is then maintained at −78° for 60 minutes and at 25° for 10 minutes. The volatile material is fractionated using baths at −23, −78, and −196°. The first trap contains traces of $(CH_3)_2GeCl_2$, identified by its ^1H NMR[2] and Raman spectra;[9] the second trap contains pure $(CH_3)_2GeHCl$ (1.2486 g, 8.98 mmole), and the third trap contains a mixture of B_2H_6 and unreacted $(CH_3)_2GeH_2$, identified by gaseous infrared spectra.[15,16] The yield of $(CH_3)_2$-GeHCl based on dimethylgermane used is about 86%.

Properties

Chlorodimethylgermane is a colorless mobile liquid (mp −74°, bp 89°). Its vapor pressure over the range −28 to +29° is given by the equation: $\log p\ (\text{torr}) = 7.33 - 1614/T$. The ^1H NMR parameters are δCH_3 (doublet) 0.80 ppm, δGeH (septet) 5.55 ppm. (Shifts downfield from TMS, 5% v/v in CCl_4.) Characteristic infrared bands are centered at 2988 (m), 2929 (m), 2083 (vs), 1255 (m), 841 (vs), 620 (s), 406 (s) cm^{-1}. in the spectrum of the gas. It is best stored at −196° in break-seal glass ampules.

C. BROMODIMETHYLGERMANE

$$(CH_3)_2GeHCl + HBr \longrightarrow (CH_3)_2GeHBr + HCl$$

■ **Caution.** *See the recommendations in the general caution on p. 155.*

Procedure

Chlorodimethylgermane (0.6644 g, 4.78 mmole) and hydrogen bromide (about 7.5 mmole) are condensed together at −196° into the reactor (see Sec. B). The mixture is allowed to warm to room temperature with occasional agitation and quenching with a bath at −78° if the reaction proceeds too vigorously. After 15 minutes at room temperature the volatile products are fractionated using baths at −78 and −196°. The former contains pure $(CH_3)_2GeHBr$ (0.8276 g, 4.51 mmole) and the latter a mixture of HCl and HBr, identified by infrared analysis. The yield of $(CH_3)_2GeHBr$ based on $(CH_3)_2GeHCl$ used is 94%.

A similar reaction using HI in place of HBr converts chlorodimethylgermane to iododimethylgermane (see Sec. D) and dichlorodimethylgermane to diiododimethylgermane.[17]

Properties

Bromodimethylgermane is a colorless liquid (extrapolated bp ~130°). Its vapor pressure over the range −20 to +23° is given by the equation: $\log p$ (torr) = 6.6 − 1500/T. In CCl_4 solution (5% v/v) the ^1H NMR chemical shifts are δCH_3 (doublet) 0.94 ppm and δGeH (septet) 5.25 ppm downfield from TMS. Principal infrared absorptions are centered at 3002 (m), 2928 (m), 2083 (s), 1422 (m), 1252 (m), 842 (vs), 768 (m), 669 (s) cm^{-1} in the spectrum of the gas. In the Raman spectrum νGe—Br is observed as a strong line at 279 cm^{-1}. Bromodimethylgermane should be stored at −196°.

D. IODODIMETHYLGERMANE

$$(CH_3)_2GeH_2 + I_2 \longrightarrow (CH_3)_2GeHI + HI$$

■ **Caution.** *See the recommendations in the general caution on. p. 155.*

Procedure

Mercury manometers should be isolated during this procedure. Iodine (1.3100 g, 5.16 mmole) is resublimed into a 10-mL ampule* which is attached to the vacuum system and thoroughly evacuated. (Occasional cooling with a bath at −78° and the use of the glass wool above the ampule will minimize iodine contamination.) The ampule is cooled to −196° and dimethylgermane (about 6.0 mmole) is condensed into it. The reactants are isolated from the vacuum system

*This is achieved conveniently using the Schlenk tube technique. See reference 8, pp. 145 ff.

and maintained at −78°. The reaction may be accelerated by occasional slight local warming with the fingers. The ampule should be open to the manifold (isolated) during this operation to provide a shock absorber in case of rapid HI evolution. The checkers recommend an alternative precaution against sudden pressure release by replacing the 10-mL ampule with a 100-mL, round-bottomed flask fitted with a freeze-out tip (see reference 8 p. 102). After about 60 minutes the mixture should appear straw colored with no residual solid iodine. The volatile material is fractionated using baths at −23, −78, and −196°. The first trap retains a small amount of $(CH_3)_2GeI_2$, identified by ^1H NMR[2] and Raman spectroscopy[17]; pure $(CH_3)_2GeHI$ (1.1433 g, 4.96 mmole) is obtained in the trap at −78°; and a mixture of $(CH_3)_2GeH_2$ and HI, identified in gas-phase infrared spectra,[15] condenses in the following trap. The yield of $(CH_3)_2GeHI$ based on the conversion of $(CH_3)_2GeH_2$ is about 83% (Checkers obtained about 94%).

Properties

Iododimethylgermane is a colorless liquid (extrapolated bp ~175°). Its vapor pressure over the range −8 to +34° is given by $\log p$ (torr) = 6.2− 1509/T. The ^1H NMR chemical shifts measured in CCl_4 solution (5% v/v) are δGeH (septet) 4.71 ppm downfield from TMS. In the infrared spectrum of the gas, principal features are centered at 2995 (m), 2926 (m), 2077 (s), 1418 (m), 1252 (m), 839 (vs), 763 (m), 654 (m), 615 (s), 591 (s) cm^{-1}. In the Raman effect the characteristic νGe−I is observed as an intense line at 231 cm^{-1}. Iododimethylgermane should be stored at −196°.

E. FLUORODIMETHYLGERMANE

$$2(CH_3)_2GeHI + PbF_2 \longrightarrow 2(CH_3)_2GeHF + PbI_2$$

■ **Caution.** *See the recommendations in the general caution on p. 155.*

Procedure

One end of a column (about 25 mm od × 300 mm) is sealed and fashioned into a bulb to permit condensation of volatile species. The other end is attached to a vacuum system. Anhydrous powdered PbF_2 (about 25 g) is packed loosely into the column using Pyrex glass wool as a support medium. The column is evacuated thoroughly for at least 30 minutes. Iododimethylgermane (0.4956 g, 2.15 mmole) is then allowed to pass back and forth through the column. An immediate intense yellow coloration, which gradually passes down the column, signifies the formation of PbI_2. After five such double passes the volatile products are fractionated using baths at −63 (chloroform slush), −126, and −196°. The first trap contains traces of $(CH_3)_2GeHI$, identified spectroscopically; pure $(CH_3)_2GeHF$ (0.2157 g, 1.76 mmole) condenses in the second trap and the third trap contains small amounts of $(CH_3)_2GeH_2$ (<0.1 mmole), identified by its ^1H NMR spectrum.[14] The yield of $(CH_3)_2GeHF$ based on the $(CH_3)_2GeHI$ used is 82%. [If the column has been packed inefficiently, the first trap may contain unreacted $(CH_3)_2GeHI$, which may be recycled. If the PbF_2 is at all hydrous the quantity of $(CH_3)_2GeH_2$ in the trap at −196° increases and this is accompanied by $(CH_3)_2GeF_2$ in the trap at −63°. The latter may be retained and characterized spectroscopically.[2,18]]

Properties

Fluorodimethylgermane is a colorless mobile liquid (mp −31°, bp 63°). The vapor pressure in the range −21 to +26° is given by $\log p$ (torr) = 7.98 − 1715/T. In the ^1H NMR spectrum of dilute solutions* or at low temperature the typical doublet-septet resonances of the other halides are duplicated owing to additional coupling with ^{19}F, namely, δCH_3 0.59 ppm, J^{vic}_{HF} 7.0 Hz; δGeH 5.79 ppm, J^{gem}_{HF} 46.8 Hz (shifts downfield from TMS, 5% v/v in CCl_4). In the infrared spectrum of the gas, strong lines are centered at 2998 (m), 2085 (s), 1424 (m), 1256 (m), 845 (s), 760 (m), 707 (m), 670 (s), 626 (s) cm^{-1}. The product disproportionates slowly at ambient temperature, particularly in the liquid phase, and should be stored at −196°.

References

1. E. A. Flood, K. L. Godfrey, and L. S. Foster, *Inorg. Synth.*, **3**, 64 (1960); M. Wieber, C. D. Frohning, and M. Schmidt, *J. Organomet. Chem.*, **6**, 427 (1966); A. P. Belij, A. I. Gorbunov, S. Golubtsov, and N. S. Feldshtein, *J. Organomet. Chem.*, **17**, 485 (1969), and references therein; V. S. Ponomarenko and G. Ya Vzenkova, *Izv. Akad. Nauk S.S.S.R.*, **1957**, 1020; R. E. J. Bichler, M. R. Booth, H. C. Clark, and B. K. Hunter, *Inorg. Synth.*, **12**, 64 (1970).
2. G. K. Barker, J. E. Drake, and R. T. Hemmings, *Can. J. Chem.*, **52**, 2622 (1974).
3. J. E. Drake, R. T. Hemmings, and C. Riddle, *J. Chem. Soc. (A)*, **1970**, 3359.
4. J. E. Griffiths, *Inorg. Chem.*, **2**, 375 (1963).
5. E. Amberger and H. Boeters, *Angew. Chem.*, **73**, 114 (1961).
6. C. H. Van Dyke, J. E. Bulkowski, and N. Viswanathan, *Inorg. Nucl. Chem. Lett.*, **7**, 1057 (1971).
7. V. F. Mironov, L. N. Kalinina, E. M. Berliner, and T. K. Gav, *Zh. Obshch. Khim.*, **40**, 2590 (1970).
8. D. F. Shriver, *The Manipulation of Air-sensitive Compounds*, McGraw-Hill Book Co., New York, 1969.

*Characteristic of germyl fluorides,[6] exchange processes lead to broadened featureless spectra in concentrated solutions.

9. J. E. Griffiths, *Spectrochim. Acta,* **20**, 1335 (1964).
10. J. R. Bates, J. O. Halford, and L. C. Anderson, *J. Chem. Phys.*, **3**, 531 (1935).
11. T. Wentink and V. H. Tiensu, *J. Chem. Phys.*, **28**, 826 (1958).
12. A. Stock and O. Priess, *Chem. Ber.*, **47**, 3109 (1914).
13. S. Cradock, *Inorg. Synth.*, **15**, 165 (1974).
14. H. Schmidbaur, *Chem. Ber.*, **97**, 1639 (1964).
15. D. F. Van de Vondel and G. P. Van der Kelen, *Bull. Soc. Chim. Belges,* **74**, 467 (1965); J. E. Griffiths, *J. Chem. Phys.*, **38**, 2879 (1963).
16. R. C. Lord and E. Nielson, *J. Chem. Phys.*, **19**, 1 (1951).
17. J. W. Anderson, G. K. Barker, J. E. Drake, and R. T. Hemmings, *Can. J. Chem.*, **49**, 3931 (1971).
18. J. W. Anderson, G. K. Barker, A. J. F. Clark, J. E. Drake, and R. T. Hemmings, *Spectrochim. Acta,* **30A**, 1081 (1974).

29. IODOGERMANE, DIGERMYLCARBODIIMIDE, DIGERMYL SULFIDE, AND THIOGERMANES

Submitted by JOHN E. DRAKE,* RAYMOND T. HEMMINGS,† and H. ERNEST HENDERSON*
Checked by ARLAN D. NORMAN‡

The interaction of germyl carbodiimides with weak protic acids, for example, CH_3OH and H_2E (E = O, Se), provides useful synthetic routes to the production of a variety of germyl derivatives, namely, GeH_3OCH_3 and $(GeH_3)_2E$.[1,2] The advantage of this type of reaction is that the only other product is a nonvolatile white solid that has been described as a polymerized cyanamide of the form $(H_2NCN)_n$.[1,3]

Digermylcarbodiimide has been prepared by the reaction of fluorogermane and bis(trimethylsilyl)carbodiimide.[3] A recent report suggests an alternative route involving the passage of iodogermane through a packed lead cyanamide column.[4] The latter method has been generalized for the preparation of methyl-substituted digermyl and disilylcarbodiimides of the formula $[(CH_3)_nH_{3-n}MN:]_2C$, M = Si, Ge, $n = 0 \rightarrow 3$, and has been found to give consistently high yields.[4]

Digermyl sulfide (digermathiane) has been prepared by the reaction of iodogermane with mercury(II) sulfide[5] and, more recently, from bromogermane and dilithium sulfide in dimethyl ether.[6] (Methylthio)germane has been prepared by

*Department of Chemistry, University of Windsor, Windsor, Ontario, Canada N9B 3P4.
†Department of Chemistry, University of the West Indies, Mona, St. Andrew, Jamaica, W.I.
‡Department of Chemistry, University of Colorado, Boulder, CO 80309.

the interaction of CH_3SNa with either GeH_3Cl or GeH_3I,[7] while (phenylthio)-germane involves the reaction of C_6H_5SK with GeH_3Br in diethyl ether.[8] We report herein the details of the preparation of digermyl sulfide, (methylthio)-germane, and (phenylthio)germane using the carbodiimide as an intermediate. In addition, details for the preparation of the precursors, iodogermane and digermylcarbodiimide, are given. The syntheses reported here give substantially higher yields in a relatively short time compared to other methods. Each reaction, including the preparation of digermylcarbodiimide, takes about 3 hours.

Starting Materials

All reactions are carried out by standard vacuum-line techniques[9] on a previously cleaned and dried vacuum line. Since germanes and their derivatives are soluble in hydrocarbon grease, Teflon-in-glass high-vacuum valves are used.

Germane*, GeH_4, is passed through traps at −126° (methylcyclohexane slush) and degassed at −196° (liquid N_2). Its purity is confirmed by its infrared spectrum.[11] Hydrogen sulfide† is prepared by the hydrolysis of Al_2S_3[12] and its purity is checked by its vapor pressure.[13] Methanethiol,‡ CH_3SH, is purified by vacuum fractionation through traps at −78° (trichloroethylene-solid CO_2) and condensation at −196°; it is characterized by its vapor pressure and infrared spectrum.[14] Benzenethiol,‡ C_6H_5SH, is degassed at −23° (CCl_4/liquid N_2 slush) and its purity is established by infrared spectroscopy.[15]

■ **Caution.** *The germanes and sulfur compounds used in these preparations should be regarded as toxic. Their exposure to air and/or moisture is likely to promote rapid decomposition. Manipulations should be carried out in a vacuum system in a well-ventilated area. Temporary lubrication of detachable ground-glass joints should be effected with a silicone-type lubricant. For the above reasons we describe small-scale preparations with pressures not to exceed 1 atm in the vacuum line.*

A. IODOGERMANE§

$$GeH_4 + I_2 \longrightarrow GeH_3I + HI$$

*GeH_4 may be obtained from Matheson Gas Products, East Rutherford, NJ or it may be prepared.[10]

†Also available from Matheson Gas Products.

‡Aldrich Chemical Co., Milwaukee, WI.

§Iodogermane may also be prepared by the reaction of chlorogermane with hydrogen iodide.[16] The method described herein was first reported in *Synth. Org. Metal-Org. Chem.*, 3, 125 (1973).

Procedure

The reactor is a thick-walled glass tube (about 200 mm long, 5 mm od, 3 mm id) with a constriction about 150 mm from the closed end and a ground-glass joint at the other. Iodine (4.0 mmole) is sublimed into the reactor, using the Schlenk tube technique.[9] The reactor is held at −78° and reevacuated. The bath at −78° is replaced by one at −196° and GeH_4 (6.0 mmole) is condensed into the reactor, which is then sealed off at the constriction, leaving a tapered end. The reaction tube is placed into a thick-walled metal tube (about 10 mm id) in a Dewar flask at −78°. The flask is allowed to warm slowly to room temperature over about 6 hours, after which time the mixture is usually dark brown. The reactor is shaken *gently* to ensure mixing (■ **Caution.** *High internal pressure*). The reaction is allowed to continue until the color of the mixture is straw-yellow, indicating that the iodine has been consumed (about 3 hr). At this time the tube is placed into a conventional ampule opener,[9] which is then attached to the vacuum line, evacuated, and *cooled thoroughly* (about 15 min at −196°). The tapered end of the tube is then broken and traces of noncondensible gas are pumped off before the products are allowed to melt. The volatile species are fractionated through a series of traps at −23, −78, and −196°. The trap at −23° retains any polyiodinated species formed and the trap at −78° retains iodogermane* (GeH_3I, 3.2 mmole, 76%); its purity may be checked by its ^1H NMR[17] and infrared[18] spectra. Unreacted germane and hydrogen iodide, formed in the reaction, condense in the trap at −196°. The GeH_4 may be recycled after passage through a trap at −126°.

For properties of GeH_3I see reference 16.

B. DIGERMYLCARBODIIMIDE

$$2GeH_3I + PbCN_2 \longrightarrow (GeH_3N{:})_2C + PbI_2$$

Procedure

A column (about 25 mm od × 300 mm long), packed loosely with glass wool and lead cyanamide (about 25 g, Alfa Inorganics Inc., Beverly, MA†), is attached to a vacuum line and evacuated for about 2 hours. Iodogermane (GeH_3I; 0.101

*Because mercury initiates rapid decomposition of iodogermanes, the manometers are isolated from the line when not in use. Iodogermane is light sensitive,[19] so exposure to direct light should be minimized.

†If not available commercially, $PbCN_2$ can be prepared according to the method of S.A. Miller and B. Bann, *J. Appl. Chem. (London)*, 6.89 (1956)

g, 0.50 mmole) is allowed to pass twice over the column by placing a bath at $-196°$ around the bottom of the column. The progress of the reaction can be monitored because the yellow $PbCN_2$ darkens as the iodogermane passes over it. The products are collected by pumping into a trap at $-196°$, whence the carbodiimide may be purified finally be allowing it to distill through traps at $-45°$ (chlorobenzene-liquid N_2 slush) and $-196°$. The former trap contains the pure $GeH_3NCNGeH_3$ (0.0402 g, 0.21 mmole; purity is confirmed by observation of its ^1H NMR spectrum[1]). The yield of digermylcarbodiimide is 84% (checker 69%).

For properties of $(GeH_3N)_2C$, see reference 3.

C. DIGERMYL SULFIDE (Digermathiane)

$$GeH_3NCNGeH_3 + H_2S \longrightarrow (GeH_3)_2S + H_2NCN$$

Procedure

Digermylcarbodiimide ($GeH_3NCNGeH_3$: 0.0459 g, 0.24 mmole) is condensed, *in vacuo*, into a previously evacuated 100-mL reaction flask (equipped with a Teflon-in-glass stopcock and ground-glass joint) by placing a bath at $-196°$ around the flask. When the transfer is complete, the stopcock is closed, the system is reevacuated, and hydrogen sulfide (H_2S; about 1.0 mmole) is condensed, in a similar manner, into the flask. The stopcock is again closed, the bath at $-196°$ removed from around the flask, and the contents allowed to warm to room temperature. The reaction of digermylcarbodiimide and hydrogen sulfide is complete after 30 minutes, at which time, the stopcock is opened and the contents are allowed to distill through U-traps at -78 and $-196°$.

The trap at $-78°$ contains pure digermyl sulfide [$(GeH_3)_2S$; 0.0404 g, 0.22 mmole; identified by its ^1H NMR, infrared, and Raman spectra[20]]. Excess hydrogen sulfide, identified by infrared spectroscopy,[14] collects in the trap at $-196°$, while a small amount of nonvolatile white material remains in the flask. The yield of digermyl sulfide, based on the carbodiimide, is 92%.

Properties

Digermyl sulfide is a colorless liquid that has a vapor pressure of 5 torr at 0°.[20] It is relatively stable at room temperature when sealed in a Pyrex glass tube, but it is recommended that it be kept at $-196°$. Its vapor pressure permits easy handling under vacuum conditions. It has a very distinct foul odor and should be kept in (or close to) a hood at all times.

The ^1H NMR spectrum consists of a singlet at 4.64 ppm shifted downfield from tetramethylsilane in cyclohexane solution. Principal features in its infrared spectrum above 400 cm^{-1} are 2110 (s) (sh), 2097 (s), 872 and 849 (m), 823 and 816 (s), 577 and 556 (m), 412 (s).[20]

D. (METHYLTHIO)GERMANE

$$GeH_3NCNGeH_3 + 2CH_3SH \longrightarrow 2GeH_3SCH_3 + H_2NCN$$

Procedure

Digermylcarbodiimide ($GeH_3NCNGeH_3$; 0.085 g, 0.445 mmole) and methanethiol (CH_3SH; about 1.0 mmole) are condensed *in vacuo* into a 100-mL reaction flask equipped with a Teflon-in-glass stopcock and a ground-glass joint as described in Section C. The reactants are isolated in the flask by closing the stopcock and then left to react at room temperature for about 30 minutes. The stopcock is then opened and the products allowed to distill through U-traps at -78 and -196°.

The trap at -78° contains pure (methylthio)germane (GeH_3SCH_3; 0.088 g, 0.718 mmole; identified by its ^1H NMR, infrared, and Raman Spectra[7,21]). Excess methanethiol, identified by infrared spectroscopy,[14] is collected in the trap at -196°. A small amount of nonvolatile white product remains in the reaction vessel. The yield of (methylthio)germane, based on digermylcarbodiimide, is 82%.

Properties

(Methylthio)germane is a volatile, colorless liquid, the vapor pressure of which obeys the equation $\log P_{(torr)} = \frac{-1537.9}{T} + 7.1555$ over the range -50.2 to +18.4°.[7] The boiling point (extrapolated) is 87° and the Trouton's constant is 19.6 cal/deg mole. The compound is fairly stable at room temperature when sealed in a Pyrex glass tube and is easy to handle under vacuum conditions, although care should be taken because of its nauseating odor.

The ^1H NMR spectrum consists of two singlets; δGeH at 4.48 ppm and δCH_3 at 2.04 ppm (both shifted downfield from tetramethylsilane in cyclohexane solution). Principal features in its infrared spectrum above 400 cm^{-1} are 2936 (s), 2113 and 2078 (vs), 873 (s), 832 (s), 567 (vs), 410 (s).[21]

E. (PHENYLTHIO)GERMANE

$$GeH_3NCNGeH_3 + 2C_6H_5SH \longrightarrow 2GeH_3SC_6H_5 + H_2NCN$$

Procedure

A small cylindrical reaction vessel (volume about 5 mL) equipped with a Teflon-in-glass stopcock and ground-glass joint is attached to the vacuum line and evacuated. The system is then isolated from the pump, flooded with nitrogen and, with the stopcock removed, benzenethiol (C_6H_5SH; 0.1355 g, 1.23 mmole) is introduced into the reaction vessel with a syringe. The stopcock is replaced, a bath at $-23°$ is placed around the vessel, and the system is reevacuated. Digermylcarbodiimide($GeH_3NCNGeH_3$; 0.1338 g, 0.70 mmole) is then condensed into the vessel at $-196°$. The stopcock is closed and the reactants are allowed to warm to room temperature. The reaction is complete in 1 hour, after which the contents are pumped through U-traps at -23 and $-196°$.

The trap at $-23°$ contains pure (phenylthio)germane ($GeH_3SC_6H_5$; 0.181 g, 0.98 mmole) which is identified by its ^1H NMR spectrum.[8] The trap at $-196°$ contains excess digermylcarbodiimide, which is identified by its ^1H NMR spectrum.[1] A small amount of nonvolatile white product remains inside the reaction vessel. The yield of (phenylthio)germane, based on benzenethiol consumed, is 80% (checker 69%).

Properties

(Phenylthio)germane is a colorless liquid with a characteristically foul odor. It is quite stable at room temperature in sealed glass ampules and may be distilled satisfactorily under high vacuum without heating. The ^1H NMR spectrum in cyclohexane solution consists of two resonances; δGeH at 4.71 ppm and δCH, a multiplet at 7.20-6.95 ppm (downfield of tetramethylsilane). Strong features in the solid-phase infrared spectrum are observed above 600 cm^{-1} at 2048 (s), 1477 (s), 1436 (ms), 1023 (ms), 864 and 846 (ms), 798 (vs), 737 (vs), 696 and 688 (vs).[8]

References

1. S. Cradock and E. A. V. Ebsworth, *J. Chem. Soc. (A)*, **1968**, 1423.
2. S. Cradock, E. A. V. Ebsworth, and D. W. H. Rankin, *J. Chem. Soc. (A)*, **1969**, 1628.
3. S. Cradock, *Inorg. Synth.*, **15**, 164 (1974).
4. J. E. Drake, R. T. Hemmings, and H. E. Henderson, *J. Chem. Soc. (Dalton)*, **1976**, 366.
5. S. Sujishi, Abstracts, 17th International Congress of Pure and Applied Chemistry, 1959, p. 53.
6. D. W. H. Rankin, *Inorg. Synth.*, **15**, 182 (1974).
7. J. T. Wang and C. H. Van Dyke, *Chem. Commun.*, **1967**, 612; *Inorg. Chem.*, **7**, 1319 (1968).
8. C. Glidewell and D. W. H. Rankin, *J. Chem. Soc. (A)*, **1969**, 753.
9. D. F. Shriver, *The Manipulation of Air-sensitive Compounds*, McGraw Hill Book Co., New York, 1969.

10. W. L. Jolly and J. E. Drake, *Inorg. Synth.*, 7, 34 (1963); A. D. Norman, J. R. Webster, and W. L. Jolly, *Inorg. Synth.*, **11**, 176 (1968).
11. J. W. Straley, C. H. Tindal, and H. H. Nielson, *Phys. Rev.*, **62**, 161 (1942).
12. A. Tian and S. Aubanel, *Compt. Rend. Trav. Fac. Sci. Mars.*, **1**, 97 (1942).
13. *Handbook of Chemistry and Physics*, 54th ed., D-184.
14. R. H. Pierson, A. N. Fletcher, and E. S. C. Gantz, *Anal. Chem.*, **28**, 1218 (1956).
15. C. J. Pouchert, *The Aldrich Library of Infrared Spectra*, Aldrich Chemical Co., Milwaukee, WI, 1970, p. 118.
16. S. Cradock, *Inorg. Synth.*, **15**, 161 (1974).
17. E. A. V. Ebsworth, S. G. Frankiss, and A. G. Robiette, *J. Mol. Spectrosc.*, **12**, 299 (1964).
18. D. E. Freeman, K. H. Rhee, and M. K. Wilson, *J. Chem. Phys.*, **39**, 2908 (1963).
19. E. Wiberg and E. Amberger, *Hydrides of the Elements of Main Groups I-IV*, Elsevier Publishing Co., New York, 1971, p. 692.
20. T. D. Goldfarb and S. Sujishi, *J. Am. Chem. Soc.*, **86**, 1679 (1964).
21. S. Cradock, *J. Chem. Soc. (A)*, **1968**, 1426.

Chapter Seven

PHOSPHORUS COMPOUNDS

30. TERTIARY PHOSPHINES

$$Mg + RX \xrightarrow{(C_2H_5)_2O} RMgX$$

$$2\ RMgX + C_6H_5PCl_2 \xrightarrow{(C_2H_5)_2O} C_6H_5PR_2 + 2\ MgClX$$

Submitted by VITO DONATO BIANCO* and SALVATORE DORONZO*
Checked by R. BRUCE KING† and W. F. MASLER†

The organic phosphines of the type $C_6H_5PR_2$ are used frequently as ligands in transition metal organometallic chemistry. It is useful to report in detail an easy method for their preparation.

The tertiary phosphine $C_6H_5PEt_2$ can be prepared by reaction of $C_6H_5PCl_2$ with aluminum trichloride and tetraethyl lead[1] and $C_6H_5PBu_2$ can be prepared by treatment of white phosphorus with phenyllithium and butyl bromide (1-bromobutane),[2] but the most useful method, which is of general applicability to a wide range of phosphines, involves the reaction between alkyl (or aryl) magnesium halides and dichlorophenylphosphine (phenylphosphonic dichloride) in diethyl ether[3-9]; it is easy to use, not time consuming,‡ and offers high yields.

*Istituto di Chimica Generale ed Inorganica, Università di Bari, Bari, Italy.
†University of Georgia, Athens, GA 30602.

‡The operative steps reported by D. M. Adams and J. B. Raynor in *Advanced Practical Inorganic Chemistry,* John Wiley, and Sons Ltd., London, 1965, p. 116 are long and require many manipulations, and the apparatus reported therein is not simple to assemble.

The phosphines reported are $C_6H_5P(C_2H_5)_2$, $C_6H_5P(n\text{-}C_4H_9)_2$, $C_6H_5P(C_6H_{11})_2$, and $C_6H_5P(CH_2C_6H_5)_2$.

■ **Caution.** *Phosphines are foul-smelling, toxic compounds. All the manipulations should be carried out in an efficient fume hood.*

Important. Since the phosphines are easily oxidized, all reactions must be carried out under an atmosphere of dry, oxygen-free nitrogen, and all reaction products must be manipulated in the absence of air.

Procedure

Diethyl ether is purified and dried in the usual way[10] or by distillation under nitrogen from lithium tetrahydridoaluminate immediately before use and storage under a nitrogen atmosphere. Magnesium turnings are washed before use by shaking them twice with diethyl ether.* Alkyl halides are distilled freshly.* Since the procedures employed are essentially the same, only the preparation of diethylphenylphosphine is reported in detail as a typical procedure for the preparation of phosphines.

A. DIETHYLPHENYLPHOSPHINE

A 1-L, three-necked, round-bottomed flask is equipped with a reflux condenser, mechanical stirrer, 250-mL pressure-equalizing dropping funnel, and T-tube, one side of which is attached to a source of dry nitrogen and the other to an oil bubbler.

To the dropping funnel is added 38 g (26 mL, 0.349 mole) of ethyl bromide in 50 mL of dry diethyl ether and the flask is charged with 8.5 g (0.349 mole) of magnesium turnings and 150 mL of dry diethyl ether. About 2 mL of C_2H_5Br solution is then added with slow stirring. After the reaction has started† the remaining ethyl bromide is added slowly over a period of 45 minutes (continuing moderate stirring). When the addition is complete, the solution of Grignard compound is stirred at room temperature for an additional 30 minutes. The dropping funnel is then charged with a solution of 28 g (21.2 mL, 0.157 mole) of $C_6H_5PCl_2$ in 60 mL of dry diethyl ether and the flask is cooled in an ice-salt bath. The solution of $C_6H_5PCl_2$ is then added dropwise over a period of about 1 hour to the Grignard reagent, with stirring. After the addition is complete, the reaction mixture is refluxed for ½ hour, cooled in an ice bath, and treated slowly

*The checkers used magnesium turnings (purchased from Fisher Scientific) as received with no washing. With the exception of the ether, all organics (purchased from Aldrich Chemical Company) were used as received with little or no reduction in yield.

†Some crystals of iodine or ethylene dibromide (1,2-dibromoethane) may be used to initiate the reaction.

with 200 mL of a saturated aqueous solution of ammonium chloride (the solution should be deoxygenated with nitrogen prior to use to prevent oxidation of the phosphines during working). At this point the reflux condenser, the mechanical stirrer, and the dropping funnel are removed (under nitrogen flow), the main neck is stoppered, and the reaction mixture is transferred (under nitrogen atmosphere) to a separatory funnel. The flask is washed with two 15-mL portions of diethyl ether and the washings are added to the mixture in the separatory funnel. The mixture is shaken vigorously, the organic layer is separated, and the aqueous layer is extracted three times with diethyl ether (80, 50, 30 mL) . The organic fractions are collected and dried over Na_2CO_3 (or $MgSO_4$) for 4 hours and are then stored under a nitrogen atmosphere. The solution is filtered from the dessiccant and the solvent is distilled at atmospheric pressure or removed at room temperature and ~30 torr.

The residue is fractionated under reduced pressure to yield 23.6 g (90.5%) of pure phosphine as a colorless liquid (bp 126-130°/30 torr).* *Anal.* Calcd. for $C_{10}H_{15}P$: C, 72.29; H, 9.04; P. 18.67. Found: C, 72.0; H, 9.1; P, 18.6.

B. DIBUTYLPHENYLPHOSPHINE

This phosphine is prepared in a similar 2-L reaction appartus by adding (over a period of 2 hours) 123.5 g (96.8 mL, 0.90 mole) of *n*-butyl bromide in 100 mL of dry diethyl ether to 22.0 g (0.90 mole) of magnesium turnings in 340 mL of dry diethyl ether. The resulting solution of *n*-butylmagnesium bromide is cooled with an ice-salt bath and a solution of 73.3 g (55.5 mL, 0.41 mole) of $C_6H_5PCl_2$ in 100 mL of dry diethyl ether is added dropwise with stirring. After the addition is complete, the mixture is refluxed for 30 minutes and then worked up, as reported above using 500 mL of a deoxygenated, saturated solution of ammonium chloride for the hydrolysis. The yield is 75 g (82.5%) of phsophine (bp 104-108°/1.7-2.0 torr). *Anal.* Calcd. for $C_{14}H_{23}P$: C, 75.67; H, 10.36; P, 13.96. Found: C, 75.70; H, 10.25; P, 13.84.

C. DICYCLOHEXYLPHENYLPHOSPHINE

The phosphine is prepared according to the above procedure starting with 24.1 g (0.99 mole) of magnesium turnings in 350 mL of dry diethyl ether and 162 g (122 mL, 0.99 mole) of cyclohexyl bromide (bromocyclohexane) in 130 mL of dry diethyl ether. The resulting Grignard reagent is cooled in an ice-salt bath and treated dropwise (2 hr) with 81 g (61.4 mL, 0.45

*The yields obtained by the checkers were as follows: $(C_2H_5)_2C_6H_5P$, 87.5%; $(n\text{-}C_4H_9)_2C_6H_5P$, 90.6%; $(C_6H_{11})_2C_6H_5P$, 61.4%; $(C_6H_5CH_2)_2C_6H_5P$, 84.0%.

mole) of $C_6H_5PCl_2$ in 100 mL of dry diethyl ether. After the addition is complete, the mixture is refluxed for 30 minutes and then hydrolyzed by adding 500 mL of a deoxygenated, saturated solution of ammonium chloride. The diethyl ether is distilled completely from the organic layer or removed at room temperature and ~30 torr. The oily crude dicyclohexylphenylphosphine crystallizes on removal of the last traces of solvent and may be recrystallized by slow cooling of a hot, oxygen-free ethanol solution of the phosphine (1 mL/g) with stirring to prevent formation of an oil. The yield is 98 g (79.5%) of phosphine as a crystalline product (white needles) mp 56-57°. *Anal.* Calcd. for $C_{18}H_{27}P$: C, 78.83; H, 9.85; P, 11.31. Found: C, 78.70; H. 10.00; P, 11.12.

D. DIBENZYLPHENYLPHOSPHINE

The preparation is similar to that of dicyclohexylphenylphosphine using 20.4 g (0.84 mole) of magnesium turnings (in 300 mL of dry diethyl ether), 106 g (96.5 mL, 0.84 mole) of benzyl chloride (α-chlorotoluene) in 100 mL of dry diethyl ether, and 68 g (51.5 mL, 0.38 mole) of $C_6H_5PCl_2$ in 80 mL of dry diethyl ether. The reaction mixture is then refluxed for 1 hour and then hydrolyzed with 400 mL of a deoxygenated, saturated aqueous solution of ammonium chloride. After the distillation of the solvent from the organic fraction (or removal at room temperature and ~30 torr), the oily residue crystallizes on removal of the last traces of solvent. It may be recrystallized from hot oxygen-free ethanol (1.7 mL/g). Yield is 95 g (86.2%) of phosphine (white crystals), mp 68-70°. *Anal.* Calcd. for $C_{20}H_{19}P$: C, 82.76; H, 6.55; P, 10.69. Found: C, 82.60; H, 6.62; P. 10.54.

Properties

Phosphines are very reactive substances. They are readily oxidized to the phosphine oxides $C_6H_5P(O)R_2$ by reaction with hydrogen peroxide or potassium permanganate. The phosphines may be characterized by means of reaction with alkyl halides to obtain the alkylphosphonium salts; thus $C_6H_5P(C_2H_5)_2$ easily reacts with C_2H_5I to give the ethiodide $[C_6H_5P(C_2H_5)_3]I$,[1] $C_6H_5P(CH_2C_6H_5)_2$ reacts with CH_3I to give the methiodide,[6] and, similarly, $C_6H_5P(C_4H_9)_2$ and $C_6H_5P(C_6H_{11})_2$ react with alkyl iodides to give the corresponding salts.[3] Diethylphenylphosphine and dibutylphenylphosphine are soluble in THF, benzene, diethyl ether, dioxane, acetone, and ethanol. Dicyclohexylphenylphosphine and dibenzylphenylphosphine are freely soluble in the above, except for acetone and ethanol in which solvents (cold) they are only moderately soluble. All the above phosphines are insoluble in water.

References

1. L. Maier, *J. Inorg. Nucl. Chem.*, **24**, 1073 (1962).
2. M. M. Rauhut and A. M. Semsel, U.S. 3,099,691; *Chem. Abstr.*, **60**, 555b (1964).
3. W. C. Davies and W. J. Jones, *J. Chem. Soc.*, **33**, 1262 (1929).
4. W. C. Davies and W. P. Waltess, *J. Chem. Soc.*, **1935**, 1786.
5. F. G. Mann and E. J. Chaplin, *J. Chem. Soc.*, **1937**, 527.
6. F. G. Mann, J. T. Millar and F. H. C. Stewart, *J. Chem. Soc.*, **1954**, 2832.
7. R. C. Cass, G. E. Coates and R. G. Hayter, *J. Chem. Soc.*, **1955**, 4007.
8. K. Issleib and H. Völker, *Chem. Ber.*, **94**, 392 (1961).
9. M. C. Browning, J. R. Mellor, D. J. Morgan, S. A. J. Pratt, L. E. Sutton and L. M. Venanzi, *J. Chem. Soc.*, **1962**, 693.
10. A. J. Vogel, *A Textbook of Practical Organic Chemistry*, 3rd Ed., Longmans, Green and Co., London, 1956, p. 163.

31. *tert*-BUTYL FLUORO PHOSPHINES AND THEIR TRANSITION METAL COMPLEXES [*cis*-[Bis-(*tert*-butyldifluorophosphine)tetracarbonylmolybdenum(0)] and *trans*-Dibromobis(di-*tert*-butylfluorophosphine)Nickel(II)]

Submitted by O. STELZER and R. SCHMUTZLER*
Checked by PATRICIA BLUM and DEVON W. MEEK†

Alkyl and aryl fluorophosphines, R_nPF_{3-n} (R = hydrocarbon group; n = 1,2) are known to undergo spontaneous redox disproportionation[1,2] in accord with

$$3R_2PF \longrightarrow R_2PPR_2 + R_2PF_3$$

$$2RPF_2 \longrightarrow \frac{1}{x}(RP)_x + RPF_4 \qquad (\text{usually } x = 5)$$

This facile disproportionation makes the study of the coordination chemistry of these valuable ligands difficult. Although fluorodimethylphosphine and difluoromethylphosphine are known to undergo disproportionation with ease, the presence of electronegative substituents on trivalent phosphorus is known to decrease the rate of disproportionation. A similar, even more pronounced, effect has been observed upon introduction of the bulky *tert*-butyl group as a substituent, and both *tert*-butyldifluorophosphine[3] and di-*tert*-butylfluorophos-

*Lehrstuhl B für Anorganische Chemie der Technischen Universität, Pockelsstrasse 4, 33 Braunschweig, Germany.
†Department of Chemistry, Ohio State University, Columbus, OH 43210.

phine[3] are completely stable with regard to disproportionation. Both compounds are thus very useful fluorophosphine ligands.

The synthesis of the fluoro phosphines is accomplished using chlorine-fluorine exchange in the corresponding chloro phosphines with sodium fluoride in sulfolane (tetrahydrothiophene 1,1-dioxide). The preparation of representative fluoro phosphine complexes with both a zero-valent and bivalent transition metal, Mo(0) and Ni(II), respectively, is also described.

A. *tert*-BUTYLDIFLUOROPHOSPHINE

$$(CH_3)_3CPCl_2 + 2NaF \xrightarrow{\text{sulfolane}} (CH_3)_3CPF_2 + 2NaCl$$

Procedure

■ **Caution.** *The toxicity of fluoro phosphines is not known, but careful handling of the compounds inside a well-ventilated hood is recommended at all times.* tert-*Butyldifluorophosphine is spontaneously flammable in air. Special care must be taken because of its high volatility.*

The reaction is conducted in a 250-mL, three-necked flask, fitted with a mechanical stirrer, a thermometer reaching close to the bottom of the flask, and a 25-cm Vigreux column to which has been connected a distillation head with multiple receiver. *tert*-Butyldichlorophosphine[4] (31.8 g, 0.2 mole) is added to a stirred suspension of NaF* (25.2 g, 0.6 mole) in 100 mL of sulfolane† over a period of 1 hour. A slow stream of nitrogen is passed through a T-tube connected to the outlet of the distillation apparatus. After the addition of the chlorophosphine, the reaction mixture is heated as follows: 45 minutes at 80°, 15 minutes at 100-180°, and 30 minutes at 180°. Distillation of the fluorophosphine commences at 80° and is complete at 180°. The product is collected in a receiver flask cooled in Dry Ice. *tert*-Butyldifluorophosphine is obtained as a colorless liquid of bp 41-44°; the yield is 20-23 g (79-91%, based on *tert*-butyldichlorophosphine‡). *Anal.* Calcd. for $C_4H_9F_2P$: P, 24.6; F, 30.2. Found: P, 24.1; F, 29.0.

tert-Butyldifluorophosphine may be stored for short periods in glass bottles at 0°; metal cylinders are recommended for long-term storage.

*Merck reagent grade sodium fluoride employed in these preparations was dried at 100-120° for 24 hours.

†Sulfolane was obtained from Shell Chemicals and distilled twice *in vacuo* over potassium hydroxide; bp 153-154° at 18 torr.

‡The checkers report a yield of 85%, on one-half of the above scale. The use of magnetic stirring, instead of mechanical stirring, was found satisfactory by the checkers in small-scale runs.

Properties

tert-Butyldifluorophosphine is sensitive to air and moisture and must be handled in, an atmosphere of nitrogen* or argon,* or in a vacuum line throughout; the latter technique is particularly appropriate.

Physical properties and spectroscopic data for *tert*-butyldifluorophosphine, $(CH_3)_3CPF_2$ are given below:

Boiling point 41-44° (reference 3)
40-41° (checkers)

NMR data
^{31}P: δ_P −231.2 ppm (ext. H_3PO_4)[3]
226.1 ppm (checkers)
^{19}F: δ_F, +111.5 ppm (int. CCl_3F)[3]
$^1J_{PF}$, 1219 Hz[3]; 1182 Hz (checkers)
^1H: δ_H, −0.96 ppm (int. $SiMe_4$)[3]
−1.00 ppm[17]
$^3J_{PH}$, 12.3 Hz[3]; 12.9 Hz[17]; 12.5 Hz (checkers)
$^4J_{FH}$, 1.8 Hz[3]; 2.0 Hz[17]; 1.5 Hz (checkers)

The simplicity of ^1H, ^{19}F, and ^{31}P NMR spectra of both free and coordinated *tert*-butylfluorophosphine ligands is particularly noteworthy, as valuable bonding information can be obtained from the NMR spectra.[6]

Because of its stability with regard to disproportionation, $t\text{-}C_4H_9PF_2$ can be employed as a useful P(III) ligand, and a variety of complexes with transition metals, both zerovalent[6] and in positive oxidation states,[7,8] as well as those with boron acceptors,[9,18] have been prepared. The preparation of a representative complex is described below.

Oxidative fluorination of $t\text{-}C_4H_9PF_2$ with antimony trifluoride gives the tetrafluorophosphorane, $t\text{-}C_4H_9PF_4$, which may serve as a precursor to other *tert*-butylphosphorane derivatives.[3]

The vibrational spectrum of $t\text{-}C_4H_9PF_2$ has been assigned.[10]

B. *cis*-[BIS(*tert*-BUTYLDIFLUOROPHOSPHINE)TETRACARBONYL-MOLYBDENUM(0)]

$$C_7H_8Mo(CO)_4 + 2t\text{-}C_4H_9PF_2 \longrightarrow cis(t\text{-}C_4H_9PF_2)_2Mo(CO)_4 + C_7H_8$$

C_7H_8 =bicyclo[2.2.1]hepta-2,5-diene(bicycloheptadiene)

*Dried by passage over P_4O_{10} dispersed on glass wool.

Procedure

The reaction is conducted in a 100-mL Schlenk tube.[12] *tert*-Butyldifluorophosphine (1.26 g, 0.01 mole), contained in a hypodermic syringe, is added to a solution of (bicycloheptadiene)tetracarbonylmolybdenum[13] (1.5 g, 0.005 mole) in 20 mL of anhydrous pentane. There is an immediate reaction, indicated by a discharge of color. After a 5-hour stirring period the solvent and the bicycloheptadiene displaced by the fluorophosphine ligand are removed by pumping and the pressure gradually decreases from 10 to 1 torr. The oily residue thus left is distilled in a short-path distillation apparatus[14] at 100° (10^{-4} torr). The complex $[t\text{-}C_4H_9PF_2]_2Mo(CO)_4$ is obtained as a colorless oil; yield 2.1-2.2 g (91-96%). *Anal.* Calcd. for $C_{12}H_{18}F_4MoO_4P_2$: C, 31.3; H, 3.9; F 16.5; P. 13.5. Found: C, 31.5; H, 4.0; F. 16.2; P. 13.4.

Properties

cis-[Bis(*tert*-butyldifluorophosphine)tetracarbonylmolybdenum(0)] is decomposed slowly upon prolonged exposure to moist air.

NMR data

^{1}H: δ_H, -1.01 ppm (int. TMS)
$^3J_{PH}$, 16.6 Hz; $^4J_{FH}$, 1.2 Hz
^{19}F: δ_F +65.4 ppm (int. CCl_3F)
$|^1J_{PF} + {}^3J_{PF}|$ 1104 Hz
^{31}P: δ_P -263.4 ppm (ext. H_3PO_4)
IR[6]: ν_{CO} 2035 (m), 1980 (m), 1957 (s), 1951 (s) (cm^{-1})

C. DI-*tert*-BUTYLFLUOROPHOSPHINE

$$[(CH_3)_3C]_2PCl + NaF \xrightarrow{\text{sulfolane}} [(CH_3)_3C]_2PF + NaCl$$

Procedure

■ **Caution.** *The toxicity of (*t*-$C_4H_9)_2PF$ is unknown. It is recommended that the compound be handled inside a well-ventilated hood. The compound may be flammable upon exposure to air.*

Using the same apparatus as in Section A, 36.1 g (0.2 mole) of di-*tert*-butylchlorophosphine[4,5] is added by means of a hypodermic syringe to a suspension of 12.6 g (0.3 mole) of NaF in 100 mL of sulfolane. The temperature of the reaction mixture, which gradually turns brown, is raised to 150-180° and stirring is continued for another 12 hours. The fluorophosphine is recovered from the reaction mixture by distillation under reduced pressure and is obtained as a colorless liquid of bp 70-72° (55 torr); yield 28-30 g (85-91%,* based on di-*tert*-butylchlorophosphine). *Anal.* Calcd. for $C_8H_{18}FP$: C, 58.5; H, 11.0; F, 11.6; P, 18.9. Found: C, 57.4; H, 10.7; F, 10.3; P. 18.1.

The fluorophosphine may be stored in glass bottles at 0° for short periods; storage in metal containers is recommended for longer periods.

Properties

Di-*tert*-butylfluorophosphine is a colorless, volatile liquid. The compound is sensitive to air and moisture and should be handled in a nitrogen† or argon† atmosphere, or in a vacuum line.

Physical properties and spectroscopic data for $(t\text{-}C_4H_9)_2PF$ are as follows:

Boiling point 72° (55 torr) or 67-70° (48 torr[3])

NMR data

^1H: δ_H, -1.03 ppm (int. TMS)
$^3J_{PH}$, 11.9 Hz[3]; 11.3 Hz[16]; 11.5 Hz (checkers)
$^4J_{FH}$, 2.1 Hz[3]; 1.7 Hz[16]; 1.5 Hz (checkers)

^{19}F: δ_F, +223.5 ppm[3] (int. CCl_3F)
$^1J_{PF}$, 848 Hz[3]; -873.6 ± 0.02 Hz[16]

^{31}P: δ_P, -210.4 ppm[3]

The infrared spectrum of $(t\text{-}C_4H_9)_2PF$ has been assigned.[11]

D. *trans*-[DIBROMOBIS(DI-*tert*-BUTYLFLUOROPHOSPHINE)NICKEL(II)]

$$NiBr_2 + 2(t\text{-}C_4H_9)_2PF \longrightarrow Br_2Ni[(t\text{-}C_4H_9)_2PF]_2$$

Procedure

The reaction is conducted in a 100-mL Schlenk tube.[12] Di-*tert*-butylfluorophosphine (1.64 g, 0.01 mole) is added from a pressure-equalizing dropping funnel to a suspension of anhydrous $NiBr_2$[15] in 50 mL of solvent, either acetone (dried over P_4O_{10}) or benzene (dried by distillation over sodium) saturated with argon or nitrogen. After stirring for 48 hours at room temperature, excess $NiBr_2$ is

*The checkers report a yield of 85%, on one-half of the above scale. The use of magnetic stirring, instead of mechanical stirring, was found satisfactory by the checkers in small-scale runs.

†Dried by passage over P_4O_{10}, dispersed on glass wool.

removed by filtration under nitrogen through a sintered-glass funnel, and the solvent is evaporated *in vacuo* (0°/10 torr). A 2.1-2.3 g (77-82%) yield of *trans*-$[NiBr_2[(t\text{-}C_4H_9)_2PF]_2]$ is obtained. The product may be recrystallized from dichloromethane or toluene at −50° using a Schlenk tube. *Anal.* Calcd. for $C_{16}H_{36}Br_2F_2NiP_2$: C, 35.1; H, 6.6; F. 6.9; P. 11.3; MW, 546.9. Found: C, 34.9; H, 6.8; F. 7.0; P. 11.3, MW (cryoscopic in benzene), 550.

Properties

trans-[Dibromobis(di-*tert*-butylfluorophosphine)nickel(II)] is a red-brown, crystalline compound that is stable upon brief exposure to air.

The trans structure is established from an x-ray crystal structure determination.[7]

Spectral data[7] for the complexes are as follows (all spectra recorded in CH_2Cl_2):

NMR[7]

^1H: broad peak at +3.5 ppm (int. CH_2Cl_2)

^{19}F: broad doublet: δ_F, −12.3 ± 0.2 ppm (int. C_6F_6); +175 ppm (int. CCl_3F)

^{31}P: δ_P, −175 ppm (ext. H_3PO_4)

UV spectrum[7]

250 (>10.000), 315 (sh, 20,000), 407 (5900), and 485 nm (1200) (ϵ in parentheses)

Hydrolysis of $NiBr_2[(t\text{-}C_4H_9)_2PF]_2$ affords the paramagnetic, pseudotetrahedral complex, $NiBr_2[t\text{-}C_4H_9)_2HPO]_2$.[8]

References

1. V. N. Kulakova, Yu. M. Zinovev, and L. Z. Soborovskii, *Zh. Obshch. Khim.*, **29**, 3957 (1959).
2. F. Seel, K. Rudolph, and R. Budenz, *Z. Anorg. Allgem. Chem.*, **341**, 196 (1964); J. F. Nixon, in *Adv. Inorg. Chem. Radiochem.*, **13**, 363 (1970); J. F. Nixon, *Endeavour*, **32**, 19 (1972).
3. M. Fild and R. Schmutzler, *J. Chem. Soc. (A)*, **1970**, 2359.
4. M. Fild, O. Stelzer, and R. Schmutzler, *Inorg. Synth.*, **14**, 4 (1973).
5. O. J. Scherer and G. Schieder, *Chem. Ber.*, **101**, 4184 (1968).
6. O. Stelzer and R. Schmutzler, *J. Chem. Soc. (A)*, **1971**, 2867.
7. W. S. Sheldrick and O. Stelzer, *J. Chem. Soc. Dalton*, **1973**, 926.
8. O. Stelzer and E. Unger, *J. Chem. Soc. Dalton*, **1973**, 1783.
9. C. Jouany, G. Jugie, J. P. Laurent, R. Schmutzler, and O. Stelzer, *J. Chim. Phys.*, **71**, 395 (1974).
10. R. R. Holmes and M. Fild, *Spectrochim. Acta*, **27A**, 1525 (1971).
11. R. R. Holmes, G. Ting-Kuo Fey, and R. H. Larkin, *Spectrochim. Acta*, **29A**, 665 (1973).

12. D. F. Shriver, *The Manipulation of Air-sensitive Compounds*, McGraw-Hill Book Co., New York, 1969, p. 147.
13. R. B. King, *Organometallic Syntheses*, Vol. 1, J. J. Eisch and R. B. King (eds.), Academic Press, Inc., New York, London, 1965, p. 124.
14. R. Maruca, *J. Chem. Educ.*, 47, 301 (1970).
15. G. Brauer, *Handbuch der präparativen anorganischen Chemie*, Vol. II, Enke Verlag, Stuttgart, 1962, p. 1326.
16. C. Schumann, H. Dreeskamp, and O. Stelzer, *Chem. Commun.*, **1970**, 619.
17. present work.
18. E. L. Lines and L. F. Centofanti, *Inorg. Chem.*, **13**, 2796 (1974).

32. (DIALKYLAMINO) FLUORO PHOSPHORANES

$$PF_5 + (R_2N)Si(CH_3)_3 \longrightarrow (R_2N)PF_4 + (CH_3)_3SiF$$

$$(R_2N)PF_4 + (R_2N)Si(CH_3)_3 \longrightarrow (R_2N)_2PF_3 + (CH_3)_3SiF$$

Submitted by MICHAEL J. C. HEWSON* and REINHARD SCHMUTZLER*
Checked by JOSEPH G. MORSE† and JAMES J. MIELCAREK†

(Dialkylamino) fluoro phosphoranes can be prepared by the direct aminolysis of fluoro phosphoranes with secondary amines.[1-4] However, besides the desired product, large quantities of ionic species are often formed, making isolation of the covalent species difficult. This class of compounds also can be prepared by the oxidation with halogen (chlorine or bromine) of the three-coordinate dialkylamino chloro or fluoro phosphines[5] to give mixed halo phosphoranes[4,5] followed by simple halogen-fluorine exchange on the latter using Group V fluorides such as arsenic trifluoride or antimony trifluoride.[5] In some cases spontaneous exchange of ligands in the halo fluoro phosphoranes occurs to give the desired fluoro phosphorane, among other products.[6,7] This is, however, a long and involved procedure. (Dimethylamino)tetrafluorophosphorane has been prepared from the reaction of bis(dimethylamino)sulfide[8] (tetramethylsulfoxylic diamide) or tris(dimethylamino)phosphine[9] (hexamethylphosphorus triamide) with phosphorus pentafluoride; however, the scope and utility of this type of reaction are limited. A clean and convenient method of synthesis of dialkylamino fluoro phosphoranes involves the cleavage reaction of the silicon-

*Lehrstuhl B für Anorganische Chemie der Technischen Universität, Pockelsstrasse 4, 33 Braunschweig, Germany.
†Department of Chemistry, Utah State University, Logan, UT 84321.

nitrogen bond of the appropriate aminosilane (silylamine) using the Lewis acid phosphorus pentafluoride.[10-15] The advantages are the ease of preparation of the silylamines, and the only by-product, fluorotrimethylsilane, can be removed as a chemically rather inert gas, leaving the desired product in high purity. This latter reaction is described in the procedures given below. Identification of the products is accomplished easily by ^{1}H, ^{19}F, and ^{31}P nuclear magnetic resonance spectroscopy.

Interest in such compounds has arisen as a result of the extensive studies being carried out on the stereochemical aspects associated with phosphorus in coordination number five. The observation of distinct axial and equatorial P—F environments for the tetrafluorophosphoranes, by NMR techniques, is possible, whereas this observation is not common to the corresponding hydrocarbon-phosphorus bonded tetrafluorophosphoranes. Further substitution at phosphorus is possible, the tetrafluorophosphoranes and trifluorophosphoranes are reactive compounds, and it is possible to substitute further F atoms for other groups. The experimental techniques described here can be applied to the preparation of other amino fluoro phosphoranes. Further, the same techniques are applicable to other reactions involving phosphorus-fluorine compounds and silicon compounds.

■ **Caution.** *All the compounds prepared are hydrolyzed readily by moisture and water to give, among other products, hydrogen fluoride; the rate of hydrolysis decreases in the order* $PF_5 > RPF_4 > R_2PF_3 > R_3PF_2$. *The other products obtained by this hydrolysis containing at least one P—F bond are possibly toxic in nature. It is recommended, therefore, that when preparing and handling these fluoro phosphoranes gloves be worn and the preparation be carried out in an adequate hood. The toxic effects of* PF_5 *are similar to those of HF. Both severely irritate the eyes and respiratory system and may cause burns to the eyes. They irritate the skin and painful burns may develop. Liquid HF causes severe painful burns on contact with all body tissures. Excess* PF_5 *or HF, produced by inadvertently hydrolyzing* PF_5 *or products, can be disposed of by venting slowly into a water-fed scrubbing tower or column in a hood. On exposure to these compounds, immediate medical attention should be obtained, and the skin should be washed with large amounts of water. The apparatus in which the compounds are prepared and distilled should be washed well with large amounts of water, thus achieving complete hydrolysis, and eventually cleaned in "chromic acid."*

The following procedure describes the preparation of (dimethylamino)trimethylsilane[16] (pentamethylsilylamine), a precursor used in the preparation of the dimethylamino fluoro phosphoranes.

A. (DIMETHYLAMINO)TRIMETHYLSILANE (Pentamethylsilylamine)

$$2(CH_3)_2NH + (CH_3)_3SiCl \longrightarrow [(CH_3)_2N]Si(CH_3)_3 + (CH_3)_2NH_2Cl$$

Procedure

The reaction is carried out in a 4-L, three-necked flask fitted with a mechanical stirrer, gas inlet tube, and reflux condenser, the open end of which is attached to a drying tube. In a typical experiment the flask is charged with 4.0 moles (434 g) of freshly distilled chlorotrimethylsilane together with 2.5 liters of sodium-dried diethyl ether or pentane. Dimethylamine is then passed into the stirred mixture through the gas inlet tube, which is immersed about 4 cm below the surface of the liquid. The flow of gas is controlled by allowing it to pass through a bubbler before entering the flask, and a second bubbler is attached to the drying tube to control the passage of gas leaving the reaction mixture. The flow rate is adjusted so that very little or none of the gas is seen to leave the reaction mixture. A slightly exothermic reaction is observed and the solvent may start to reflux. Cooling may be necessary and is achieved by use of an ice-water bath. The addition takes 6-12 hours, and completion is noted when the rate of loss of amine is the same as the rate of addition.

Although this procedure appears to be wasteful of amine, we have no evidence that there is a higher than stoichiometric consumption of $(CH_3)_2NH$ if the reaction is conducted as described. Alternately, the $(CH_3)_2NH$ can be weighed into a suitable glass vessel and passed into the solution of the chlorosilane, using a molar ratio $(CH_3)_3SiCl/(CH_3)_2NH$ of about 1:2.1.

On completion of the addition, the reaction mixture is refluxed for about 3 hours, removing excess amine. After it is cooled, the mixture is filtered quickly, if possible under an atmosphere of dry nitrogen. The precipitate of dimethylamine hydrochloride is washed with solvent and the washings are added to the filtrate. The filtrate is distilled through a 30-cm Vigreux column at atmospheric pressure to remove the solvent and then transfered to a 500-mL flask and further distilled at atmospheric pressure. The first fraction is the remaining solvent, followed by any unreacted chlorotrimethylsilane, bp 57.7°. The second fraction is the desired product. Distillation should be slow and the bath temperature not too high, as the hydrolysis product hexamethyldisiloxane, which might be formed, bp 100°, may distil with the desired product.

(Dimethylamino)trimethylsilane is obtained as a colorless liquid of bp 86.5°/760 torr. Yields vary between 60 and 70%.

Characterization is carried out by means of ^1H NMR. Singlets appear for the methyl protons of the silyl group (-0.11 ppm) and for the methyl protons of the amine group (-2.55 ppm). The spectrum is measured using a neat liquid with benzene as an internal standard. Shifts relative to tetramethylsilane are calculated using the factor, δ benzene = -7.37 ppm relative to TMS.

B. (DIMETHYLAMINO)TETRAFLUOROPHOSPHORANE

$$PF_5 + [(CH_3)_2N]Si(CH_3)_3 \longrightarrow [(CH_3)_2N]PF_4 + (CH_3)_3SiF$$

Procedure

The reaction is carried out using a simple glass vacuum line provided with at least three connections for attaching traps and reaction tubes. A phosphorus pentafluoride cylinder* is connected to the vacuum line with standard poly(vinyl chloride) (PVC) vacuum tubing. The vacuum line also is fitted with a small detachable trap of about 50-mL volume (to enable weighing of gaseous reactants) and a simple mercury U-manometer (one arm of which is closed when evacuated) to check the pressure of the PF_5 in the system. A thick-walled glass Carius tube (about 200 mL), with a constriction to enable easy sealing, is used to carry out the reaction and is fitted with a joint to connect it to the vacuum line (Figure 4).

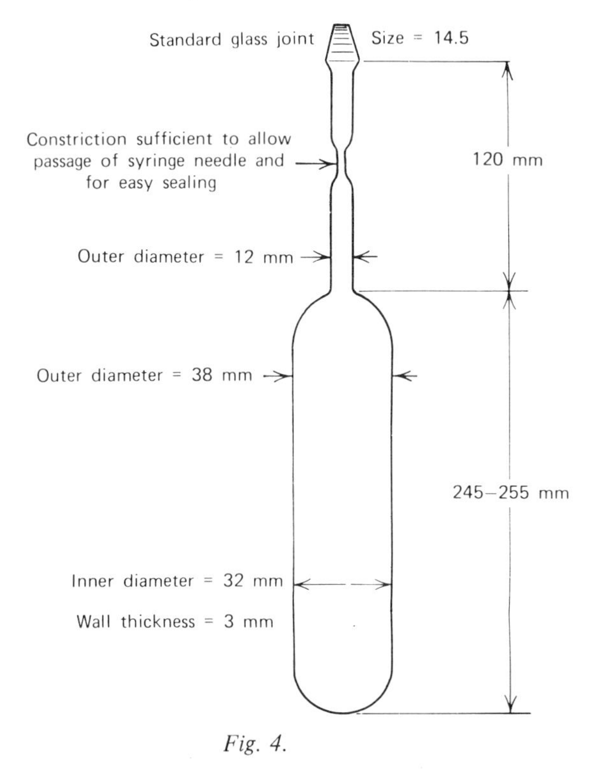

Fig. 4.

The whole system is evacuated, flame-heated and left pumping for a period of about 2 hours before the experiment is started. The vacuum line is closed to the pump when the pressure remains constant at about 0.05 torr or less and phosphorus pentafluoride is condensed directly into the trap cooled to liquid

*Phosphorus pentafluoride is used as obtained from commercial sources, such as U.S. Agrichemicals, Decatur, GA, and Ozark Mahoning Co., Tulsa, OK.

nitrogen temperature, -196°. In a typical experiment* 0.25 mole of PF_5 (31.6 g) is employed. At a pressure of about 100 torr a period of about 5 minutes is needed to condense the requisite amount into the trap. The PF_5 cylinder is then closed. The trap containing the PF_5 is closed, removed from the vacuum line, weighed, and then immediately replaced on the vacuum line and kept at -196°. The weight of PF_5 used can, however, be detemined in a slightly different manner, which is perhaps better than that given above. A Monel cylinder from the Hoke Company fitted with a suitable safety valve is employed. An advantage is that such cylinders can be cooled down to liquid nitrogen temperature (under no circumstances should normal metal cylinders be employed). The PF_5 is condensed into a previously weighed cylinder in a manner described already, closed, removed from the vacuum line, and reweighed, and the weight of PF_5 is calculated. The cylinder need not necessarily be recooled if one with a suitable safety valve is chosen.

Previously prepared (dimethylamino)trimethylsilane (pentamethylsilylamine; 0.25 mole, 29.3 g) is placed in the Carius tube, using a glass or plastic syringe, under an atmosphere of dry nitrogen. The tube is then attached to the vacuum line and slowly cooled to -196° while being pumped. When the tube is completely evacuated the vacuum line to the pump is closed, the PF_5 trap is opened and allowed to warm, and its contents are condensed slowly into the Carius tube. This procedure takes about 20 minutes. The Carius tube and contents (still at -196°) are pumped for a few minutes to ensure complete evacuation and the tube is sealed[17] at the constriction. The Carius tube is immersed in a Dewar vessel containing a Dry Ice/acetone mixture at -78°. The tube is left standing for 2 days or until the temperature rises to approximately $+10^\circ$, in some cases even higher. There is no control of the temperature during this time. The tube is then attached to the vacuum line using a piece of PVC vacuum tubing and cooled to -196° and the top is broken off carefully. After the tube has warmed to room temperature, the contents are distilled into a 100-mL flask. Any solid residue remaining in the tube is discarded. The flask is then removed from the vacuum line under a countercurrent of dry nitrogen and allowed to warm to room temperature. The contents are distilled at atmospheric pressure, using a 12-in. Vigreux column, the outlet of which is connected to a silica gel or calcium chloride drying tube. The by-product; fluorotrimethylsilane, distills first; the rate of distillation is kept as low as possible to avoid transfer of the desired product with the fluorosilane. The second fraction is the desired product; the whole process takes about ½ hour. The bath temperature is kept below 120°, thus avoiding the formation of ionic species.

*We have carried out reactions to a maximum of 0.3 mole. See reference 17 for general vacuum-line procedures. The checkers reduced the scale of reactants to one-tenth of those specified. They measured the quantity of PF_5 by measuring the gas volume on the vacuum line.

(Dimethylamino)tetrafluorophosphorane is obtained as a colorless liquid, bp 64°/760 torr. Yields vary between 70 and 85%. *Anal.* Calcd. for $C_2H_6F_4NP$: C, 15.9; H, 4.0; F, 50.3; N, 9.4; P. 20.5. Found: C, 15.8; H, 4.1; F, 48.4; N, 9.4; P, 19.3%.

The experiment can be shortened somewhat if the reaction tube is allowed to warm from -196° to room temperature over a 1-hour period. The glass tube should be enclosed in a thick walled cylindrical metal container, and placed in a hood, using explosion-proof shielding. We have, however, had no cases in which the tube has exploded. After completion of the reaction, the tube is opened and the contents are decanted directly into the distillation flask. A lower yield is obtained by this procedure because of the formation of ionic products.

Properties

(Dimethylamino)tetrafluorophosphorane is a colorless liquid that fumes when exposed to moist air. Melting points of -80[11] and -78°,[1,2] and boiling points of 60[1,2] and 63.7°,[11] have been reported. The vapor pressure[1,2,11] may be expressed by the equation

$$\log P_{(\text{torr})} = 8.12 - \frac{1764.75}{T}$$

The fluoro phosphoranes may be stored in glass bottles for brief periods, but attack is appreciable, and the use of Teflon or stainless steel containers is preferable.

The nuclear magnetic resonance spectrum consists of a single resonance for ^{19}F, split into a doublet by coupling with ^{31}P; a single resonance for the ^{31}P split into a quintet by coupling with four magnetically equivalent fluorine atoms, and a single resonance for ^{1}H split into a doublet of quintets due to coupling with ^{31}P and ^{19}F.

NMR parameters: δF = +66.8 ppm[1,2,11,12]; $^{1}J_{PF}$ = 847 Hz; δP = +69.7 ppm[4,12]; δH = -2.85 ppm[1,2,8,11,12]; $^{3}J_{HP}$ = 11.6 Hz; $^{4}J_{HF}$ = 2.05 Hz.

At -90° two distinct fluorine atom environments are observed, the structure being interpreted in terms of a trigonal bipyramid in which two fluorine atoms occupy axial sites. Two ^{19}F resonances are observed, an axial resonance, consisting of a doublet of triplets, due to ^{31}P and ^{19}F coupling, and an equatorial resonance, consisting of a doublet of triplets due to ^{31}P and ^{19}F coupling.

Low-temperature ^{19}F NMR data: $\delta F(ax)$ = +59.0 ppm,[4,12] $^{1}J_{PF(ax)}$ = 778 Hz; $^{2}J_{F(ax)F(eq)}$ = 44 Hz; $\delta F(eq)$ = +75.9 ppm $^{1}J_{PF(eq)}$ = 915 Hz; $^{2}J_{F(eq)F(ax)}$ = 44 Hz.

The low-temperature ^{31}P spectrum of $[(CH_3)_2N]PF_4$ has also been investigated[18,19] and confirms the presence of two distinct fluorine atom environments. A set of three triplets is observed.

All chemical shifts cited refer to tetramethylsilane for ^1H, trichlorofluoromethane for ^{19}F, and 85% phosphoric acid for ^{31}P. The abbreviations ax and eq refer to axial and equatorial, respectively.

Infrared spectral data have been recorded[1-3,8,11] for $[(CH_3)_2N]PF_4$. Four bands at 950, 882 (two), and 701 cm^{-1} have been assigned to P—F modes.

The photoelectron spectrum[20] (PES) of $[(CH_3)_2N]PF_4$ exhibits a single peak at 10.35 eV and is assigned unambiguously to ionization of the N(2p) lone-pair electrons. Comparison of PES data of several P—N bonded compounds supports the implication that the nitrogen lone pair is orthogonal to the axial P—F bonds.[20]

Mass spectral data have also been reported.[21] The most abundant fragment is the PF_4^+ species at *m/e* 107, characteristic of several differing tetrafluorophosphoranes.[21] The parent ion at *m/e* 151 appears in 22.9% abundance.[21] Other prominent fragments in the spectrum are PF^+, PF_2^+ and PF_3^+ caused by the successive loss of F from the PF_4^+ ion.

C. (DIETHYLAMINO)TETRAFLUOROPHOSPHORANE

$$PF_5 + [(C_2H_5)_2N]Si(CH_3)_3 \longrightarrow [(C_2H_5)_2N]PF_4 + (CH_3)_3SiF$$

Procedure

The apparatus and procedure employed for the preparation of (diethylamino)-tetrafluorophosphorane are identical to that described for the preparation of (dimethylamino)tetrafluorophosphorane (Sec. B). All of the precautions described in preparation B should be observed. Phosphorus pentafluoride (31.6 g, 0.25 mole) and (diethylamino)trimethylsilane (*N,N*-diethyl-1,1,1-trimethylsilylamine)[22] (0.25 mole, 36.3 g)* are used.

(Diethylamino)tetrafluorophosphorane is obtained as a colorless liquid of bp 99-100°/760 torr. Yields vary between 50 and 60%. *Anal.* Calc. for $C_4H_{10}F_4NP$: C, 26.8; H, 5.6; F. 42.5; N, 7.8; P, 17.3. Found: C, 27.0; H, 5.5; F, 41.6; N, 7.9; P. 17.5%. The shortened procedure from B can be used, but this gives a lower yield.

Properties

(Diethylamino)tetrafluorophosphorane is a colorless liquid fuming on contact with moist air. Its melting point is −73°[1,2] and boiling points of 99-100°[10,23] and 45-47°/150 torr[23] have been reported; $n_D^{20} = 1.3440$[24] and $d^{21} = 1.2368$[24]; the vapor pressure at 19° is 25 torr.[1,2]

*The checkers reduced the scale to one-tenth of these quantities.

The nuclear magnetic resonance spectrum is similar to that of (dimethylamino)tetrafluorophosphorane.

NMR parameters: δF = +66.5 ppm[1,2,10,12,24]; $^1J_{PF}$ = 851 Hz; δP = +69.8 ppm[10]; δH_{CH_3} = −1.21 ppm[12] δH_{CH_2} = −3.25 ppm[12]; $^3J_{HP}$ = 16-17 Hz; $^4J_{HF}$ = 1.7 Hz; $^3J_{HH}$ = 0.7 Hz.

Low-temperature ^{19}F NMR data[4,25] $\delta F(ax)$ = +60.2 ppm $^1J_{PF(ax)}$ = 793 Hz; $^2J_{F(ax)F(eq)}$ = 70 Hz; $\delta F(eq)$ = +72.3 ppm $^1J_{PF(eq)}$ = 916 Hz; $^2J_{F(eq)F(ax)}$ = 70 Hz.

D. BIS(DIMETHYLAMINO)TRIFLUOROPHOSPHORANE

$$[(CH_3)_2N]PF_4 + [(CH_3)_2N]Si(CH_3)_3 \longrightarrow [(CH_3)_2N]_2PF_3 + (CH_3)_3SiF$$

Procedure

In a typical experiment,* 0.1 mole (15.1 g) of (dimethylamino)tetrafluorophosphorane and 0.1 mole (11.7 g) of (dimethylamino)trimethylsilane (pentamethylsilylamine) are allowed to react. Both reactants are charged into a dry thick-walled glass Carius tube of about 200-mL volume, under an atmosphere of dry nitrogen. The tube is then attached to a vacuum line, cooled slowly to −196° while being pumped carefully, sealed, and allowed to warm to room temperature. The tube and contents are heated slowly to 150-170° with an oil bath and kept at this temperature for 6 hours. The reaction tube is then allowed to cool to room temperature, cooled to −196°, and opened, and a drying tube is attached to the open end. A steady flow of dry nitrogen is passed through a T-tube connected to the outlet of the drying tube. After they are warmed to room temperature, the contents of the flask are poured into a 100-mL flask under an atmosphere of dry nitrogen. Distillation then follows, using a 6-in. Vigreux column. The apparatus is subjected slowly to vacuum. The distillation takes about ½ hour. The first fraction after removal of fluorotrimethylsilane is the desired product.

Bis(dimethylamino)trifluorophosphorane is obtained as a colorless liquid of bp 49.5°/8 torr. Yields vary between 56 and 75%. *Anal.* Calcd. for $C_4H_{12}F_3N_2P$: C, 27.2; H, 6.9; F, 32.4; N, 15.9; P, 17.6. Found: C, 27.4; H, 7.0; F, 31.9; N, 16.1; P, 17.3%.

Properties

Bis(dimethylamino)trifluorophosphorane is a colorless liquid. A melting point of

*Reaction has been tried only to a maximum of 0.13 mole. The checkers obtained 47% yield using one-tenth of the quantities specified here.

$-22°$[1,2] and a boiling point of 43°/19 torr[4,12] have been reported; n_D^{23} = 1.3837.[12]

Hydrolysis it is not as rapid as for the corresponding tetrafluorophosphorane.

The nuclear magnetic resonance spectrum consists of two resonances for the ^{19}F atoms, one for the two axial fluorine atoms, split into a doublet of doublets by coupling to ^{31}P and one equatorial fluorine atom, and one for the equatorial fluorine atom, split into a doublet of triplets by coupling to ^{31}P and the axial fluorine atoms. Evidence is also seen for J_{HF} coupling in the axial resonance, although it is not well resolved. A single resonance occurs for ^{31}P, split into a doublet of triplets owing to coupling with two axial fluorine atoms and one equatorial fluorine atom. A single ^{1}H resonance is observed, split into a doublet of doublets and each of the resonance lines is further split into a triplet owing to coupling with ^{19}F and ^{31}P.

NMR parameters: δF(ax) = +54.0 ppm[1,2,4,12]; $^1J_{PF(ax)}$ = 744 Hz; $^2J_{F(ax)F(eq)}$ = 44 Hz; δF(eq) = +73.0 ppm[1,2,4,12]; $^1J_{PF(eq)}$ = 862 Hz; $^2J_{F(eq)F(ax)}$ = 44 Hz; δP = +65.0 ppm[12]; δH = −2.72 ppm[12] $^3J_{HP}$ = 10.8 Hz; $^4J_{HF(ax)}$ = 2.7 Hz; $^4J_{HF(eq)}$ = 1.5 Hz.

The photoelectron spectrum of $[(CH_3)_2N]_2PF_3$[20] exhibits peaks at 8.84 and 9.95 eV that are assigned to ionization of N(2p) lone-pair electrons.

Mass spectral data have also been reported.[21] The most abundant fragment is the $[(CH_3)_2N]PF_3^+$ species at m/e 132. The parent ion at m/e 176 appears in 4.5% abundance. Another significant ion, formed by the loss of fluorine from the parent ion, is observed at m/e 157 in 4.7% abundance.

E. BIS(DIETHYLAMINO)TRIFLUOROPHOSPHORANE

$$[(C_2H_5)_2N]PF_4 + [(C_2H_5)_2N]Si(CH_3)_3 \longrightarrow [(C_2H_5)_2N]_2PF_3 + (CH_3)_3SiF$$

Procedure

The reaction, using 0.1 mole (17.9 g) of (diethylamino)tetrafluorophosphorane* and 0.1 mole (14.5 g) of (diethylamino)trimethylsilane[22] (N,N-diethyl-1,1,1-trimethylsilylamine),* is carried out as described in Section D.

Bis(diethylamino)trifluorophosphorane is obtained as a colorless liquid, bp 69.5°/9 torr. Yields vary between 70 and 85%. *Anal.* Calcd. for $C_8H_{20}F_3N_2P$: C, 41.3; H, 8.7; F, 24.5; N, 12.1; P. 13.3. Found: C, 41.3; H, 8.7; F, 24.7; N, 12.3; P, 12.9%.

*The checkers obtained 61% yield using one-tenth of the quantities specified here.

Properties

The melting point of bis(diethylamino)trifluorophosphorane is +5°.[1,2] Boiling points of 79°/14 torr[10,12,23] and 84-86°/18 torr[23] have been reported. n_D^{25} = 1.4049.[12]

Hydrolysis occurs on contact with water.

NMR parameters: δF(ax) = +59.5 ppm[1,2,10,12,25,26]; $^1J_{PF(ax)}$ = 751 Hz; $^2J_{F(ax)F(eq)}$ = 44 Hz; δF(eq) = +67.5 ppm[1,2,10,12,25,26]; $^1J_{PF(eq)}$ = 875 Hz; $^2J_{F(eq)F(ax)}$ = 44 Hz; δP = +64.7 ppm[12,26] δH_{CH_3} = -1.14 ppm[12,26]; δH_{CH_2} = -3.10 ppm[12,26]; $^3J_{HP}$ = ~16 Hz; $^4J_{HF}$ apparent but not well resolved; $^3J_{HH}$ = 7.0 Hz.

References

1. D. W. A. Sharp, D. H. Brown, and G. W. Fraser, *Chem. Ind. (London)*, **1964**, 367.
2. D. W. A. Sharp, D. H. Brown, and G. W. Fraser, *J. Chem. Soc. (A)*, **1966**, 171.
3. G. W. Fraser, Ph.D. Thesis, University of Strathclyde, Glasgow, 1965.
4. R. Schmutzler, in *Halogen Chemistry*, Vol. 2, V. Gutmann (ed.), Academic Press, Inc., London and New York, 1967, p. 31.
5. J. Morse, K. Cohn, R. W. Rudolph, and R. W. Parry, *Inorg. Synth.*, **10**, 147 (1967).
6. J. F. Nixon, in *Advances in Inorganic Chemistry and Radiochemistry*, H. J. Emeleus and A. G. Sharpe, eds., Academic Press, London and New York, Vol. 13, 1970, p. 363.
7. G. I. Drozd, M. A. Sokal'skii, O. G. Strukov and S. Z. Ivin, *Zhur. Obshch. Khim.*, **40**, 2396 (1970).
8. D. H. Brown, K. D. Crosbie, J. F. Darragh, D. S. Ross, and D. W. A. Sharp, *J. Chem. Soc. (A)*, **1970**, 914.
9. D. H. Brown, K. D. Crosbie, G. W. Fraser, and D. W. A. Sharp, *J. Chem. Soc. (A)*, **1969**, 872.
10. R. Schmutzler, *Angew. Chem.*, **76**, 893 (1964); *Angew. Chem. Int. Ed. Engl.*, **3**, 753 (1964).
11. G. C. Demitras and A. G. McDiarmid, *Inorg. Chem.*, **6**, 1903 (1967).
12. R. Schmutzler, *J. Chem. Soc. Dalton*, **1973**, 2687.
13. M. J. C. Hewson, S. C. Peake, and R. Schmutzler, *Chem. Commun.*, **1971**, 1454.
14. M. J. C. Hewson and R. Schmutzler, *Z. Naturforsch.*, **27b**, 879 (1972).
15. M. J. C. Hewson, S. C. Peake, and R. Schmutzler, unpublished results.
16. O. Mjörne, *Svensk Kem. Tidskr.*, **62**, 120 (1950).
17. D.F. Shriver, *The Manipulation of Air-sensitive Compounds*, McGraw Hill Book Co., New York, 1969.
18. G. M. Whitesides and H. L. Mitchell, *J. Am. Chem. Soc.*, **91**, 5384 (1969).
19. M. Eisenhut, H. L. Mitchell, D. D. Traficante, R. J. Kaufmann, J. M. Deutch, and G. M. Whitesides, *J. Am. Chem. Soc.*, **96**, 5385 (1974).
20. A. H. Cowley, M. J. S. Dewar, D. W. Goodman, and J. R. Schweiger, *J. Am. Chem. Soc.*, **95**, 6506 (1973).
21. T. A. Blazer, R. Schmutzler, and I. K. Gregor, *Z. Naturforsch*, **24b**, 1081 (1969).
22. R. O. Sauer and R. H. Hasek, *J. Am. Chem. Soc.*, **68**, 241 (1946).
23. R. Schmutzler, U.S. Patent 3,300,503 (to E. I. Dupont de Nemours and Co. Inc.) (1967).

24. G. I. Drozd, M. A. Sokal'skii, V. V. Sheluchenko, M. A. Landau, and S. Z. Ivin, *Zhur. Obshch. Khim.*, 39, 935 (1969).
25. E. L. Muetterties, W. Mahler, K. J. Packer, and R. Schmutzler, *Inorg. Chem.*, 3, 1298 (1964).
26. R. Schmutzler and G. S. Reddy, *Inorg. Chem.*, 4, 191 (1965).

33. 4-(ETHYLPHENYLPHOSPHINO)-1-BUTANOL AND 1-ETHYL-1-PHENYLPHOSPHOLANIUM PERCHLORATE

Submitted by W. RONALD PURDUM,* S. D. VENKATARAMU,* and K. DARRELL BERLIN*
Checked by ROBERT R. HOLMES† and RICHARD K. BROWN†

Phospholanium salts have been of considerable interest in the last two decades with respect to the stereochemistry of the interconversion of these salts to phospholane oxides and phospholanes.[1] These investigations have centered mainly on understanding the steric requirements of phosphorus in a five-membered heterocyclic system with respect to the process of "Berry pseudorotation" and the existence of pentacoordinated phosphorus intermediates.[1,2]

Historically, phospholanium salts have been prepared by quaternization in high yield of a selected phospholane with an alkyl halide.[3] Although numerous methods exist for the preparation of the desired phospholanes (recently reviewed[4]), most of these have major drawbacks, including (1) very critical conditions, such as reaction time, dilution effects, and temperature; (2) expensive and/or difficult to manipulate reagents; and (3) the necessity of a multistep synthetic sequence giving overall low yields of phospholane. A general example of the latter would be the reaction of a substituted dihalophosphine with a 1,3-diene and hydrolysis to the phospholene oxide,[5] which then can be catalytically reduced to the phospholane oxide[6] and subsequently converted to the phospholane.[4,6]

The present procedure starts with ethyldiphenylphosphine, which is cleaved by lithium in tetrahydrofuran to lithium ethylphenylphosphide. This phosphide under vigorous reflux cleaves tetrahydrofuran to afford a necessary precursor, 4-(ethylphenylphosphino)-1-butanol (ethyl(4-hydroxybutyl)phenylphosphine). This (hydroxyalkyl)phosphine then is cyclized intramolecularly to give 1-ethyl-1-phenylphospholanium perchlorate.

The cleavage of a cyclic ether, such as tetrahydrofuran, by lithium alkylarylphosphides to afford alkylaryl(4-hydroxybutyl)phosphines alleviates a major

*Department of Chemistry, Oklahoma State University, Stillwater, OK 74074.
†Department of Chemistry, University of Massachusetts, Amherst, MA 01003.

portion of procedural difficulties and affords a stable precursor (except to oxidation) for the preparation of phospholanium salts in high yield in two steps. Besides tetrahydrofuran, this cleavage of ethers by alkali organophosphides has also been reported for the preparation of other (hydroxyalkyl)phosphines.[7,8] The technique of an intramolecular cyclization of a (bromoalkyl)phosphonium hydrobromide to a cyclic phosphonium salt is also known, but these reports have usually been applied to the synthesis of much more rigid systems, such as phosphinolinium and isophosphinolinium salts.[3,9]

The preparation of ethyldiphenylphosphine has been reported[10] and the compound is available commercially. The total working time for preparations A and B is approximately 6½ days. Utilization of this procedure has been found to be of general applicability to other phospholanium salts.[11]

A. 4-(ETHYLPHENYLPHOSPHINO)-1-BUTANOL (Ethyl(4-hydroxybutyl)phenylphosphine)

$$(C_6H_5)_2PCH_2CH_3 + 2Li + \text{THF (O)} \xrightarrow[2.\ (CH_3)_3CCl]{1.\ \text{reflux}}$$

$$C_6H_5(CH_3CH_2)P(CH_2)_4OH + C_6H_6 + LiCl$$

■ **Caution.** *In common with other alkyldiarylphosphines, ethyldiphenylphosphine presents some hazard because of its toxicity and susceptibility to air oxidation. Therefore, all operations must be performed in an efficient hood with manipulations conducted under an inert atmosphere. All ground-glass joints should be greased thoroughly with a silicone lubricant.*

Procedure

A 500-mL, three-necked flask is equipped with a mechanical stirrer, a pressure-equalizing dropping funnel, and a reflux condenser, the entire system having been previously flushed with nitrogen. To the flask is added (under nitrogen) 1.4 g (0.2 mole) of lithium shavings,* 21.4 g (0.10 mole) of ethyldiphenylphosphine,† and 200 mL of anhydrous tetrahydrofuran.‡ After it is stirred for 1

*Lithium ribbon (12.7 × 1.5 mm, 99.9%) obtained from Ventron Corporation (Alfa Products) was cut into thin strips after weighing directly into the tetrahydrofuran.

†Available from Strem Chemical Company, Danvers, MA 01923.

‡Mallinckrodt reagent grade tetrahydrofuran was distilled freshly from NaH directly before use. See *Inorg. Synth.*, **12**, 317 (1970) for precautions in handling tetrahydrofuran.

hour at room temperature, the dark-red mixture is boiled vigorously for 12 hours with the disappearance of the lithium shavings. Upon cooling to room temperature, 9.3 g (0.10 mole) of *tert*-butyl chloride is added through the addition funnel and the mixture again is heated at reflux for 24 hours. The addition of *t*-BuCl is needed to quench the phenyllithium by-product.[12] The reddish-orange mixture then is cooled to room temperature and hydrolyzed by the addition (through an addition funnel) of 5.4 g (0.10 mole) of ammonium chloride in 50 mL of water, and the resulting layers are separated. The presence of ammonium chloride facilitates hydrolysis and sufficiently increases the ionic strength of the water layer so that two layers will form. If, however, two layers do not form, solid NaCl may be added to saturate the aqueous layer and thereby facilitate separation. The aqueous layer is extracted with benzene (3 × 150 mL), the original layer and combined organic extracts are dried ($MgSO_4$), and the solvents are removed on a rotary evaporator. Distillation of the residual oil through a short-path column (Vigreux 0.5 × 4.0 in.) affords 16.0-18.0 g (76-86%) of pure ethyl(4-hydroxybutyl)phenylphosphine, bp 121-124° (0.15 torr). The product, as isolated, is pure enough for conversion to 1-ethyl-1-phenylphospholanium perchlorate if manipulated under nitrogen.

Properties

Ethyl(4-hydroxybutyl)phenylphosphine is a pale-yellow liquid slightly soluble in $CHCl_3$ and susceptible to air oxidation. As a result, it should be stored under nitrogen until used in the following preparation. Infrared absorption maxima (thin film) occur at 3.05, 3.31, 3.41, 6.32 (vw), 6.90, 7.06, 7.29, 8.03, 9.52, 11.36, 12.14, 13.64, and 14.39 μ. The proton NMR spectrum ($DCCl_3$, TMS internal standard) shows the following resonances: δ0.98 (doublet of triplets, J_{PCCH} = 16 Hz, J_{HCCH} = 7 Hz, PCH_2CH_3, 3H), 1.52 (multiplet, $CH_3CH_2PCH_2$-$CH_2CH_2CH_2OH$), 8H), 3.20 (broad singlet, O*H*, 1H), 3.50 (triplet, CH_2OH, 2H), and 7.2-7.6 (multiplet, Ar−*H*, 5H).

B. 1-ETHYL-1-PHENYLPHOSPHOLANIUM PERCHLORATE

$$C_6H_5(CH_3CH_2)P[(CH_2)_4OH] \xrightarrow[\text{reflux}]{\text{1. HBr, } C_6H_6}$$

2. $NaHCO_3$ Na_2CO_3 $HCCl_3$, H_2O
3. $NaClO_4$

$$\left[\text{(phospholane ring) } P(C_6H_5)(CH_2CH_3) \right]^+ \quad ClO_4^-$$

■ **Caution.** *In view of the well-known shock sensitivity of organic perchlorates, 1-ethyl-1-phenylphospholanium perchlorate should be handled with care. Attempts to detonate 1-6 mg samples with a hammer on a concrete surface failed to produce an explosion. It is quite probable that salts with the* PF_6^- *ion can be produced by using* KPF_6 *in place of* $NaClO_4$*, but this is not done in the present example.*

Procedure

To a nitrogen-flushed 200-mL, three-necked flask equipped with a magnetic stirrer, Dean-Stark trap[13] (and condenser), and a gas dispersion tube (fritted-glass tip) is added 75 mL of anhydrous benzene, which is then saturated with anhydrous hydrogen bromide gas (Matheson Gas Products) by bubbling at the rate of 20 mL/min for 10 minutes. This clear solution is then treated with 5.65 g (0.027 mole) of ethyl(4-hydroxybutyl)phenylphosphine in 50 mL of anhydrous benzene resulting in immediate formation of a white slurry. (If the solution acquires moisture, an oil sometimes forms at this point. However, this mixture can be used in the next step.) The mixture is heated to a vigorous reflux for 12 hours with the continuous removal of water. At the end of the time the mixture is cooled and resaturated with hydrogen bromide. An oil may sometimes form here also. This mixture is heated again to reflux for 12 hours. The benzene is distilled from the vessel to leave about 25 mL of a slurry. (On occasion an oil results. When allowed to cool, the oil solidifies. This solid can then be washed with hexane. If a slurry results, as is usual, the following procedure is used.) Upon cooling to room temperature, the residual slurry is triturated with 75 mL of hexane, resulting in the formation of a white precipitate. The solid (tentatively considered to be the 4-(bromobutyl)ethylphenylphosphine hydrobromide) is observed to be very hygroscopic and becomes an oily residue if allowed to stand in air. It is collected by a rapid vacuum filtration (sintered-glass funnel, aspirator), dissolved in 75 mL of chloroform, and extracted for 15 minutes under nitrogen with 75 mL of an aqueous 5% sodium hydrogen carbonate solution containing 5 g of disodium carbonate. The layers are separated and the aqueous layer is extracted with chloroform (2 × 100 mL). The aqueous layer is stoppered and set aside at room temperature for 48 hours. After they are dried ($MgSO_4$) for 1 hour, the combined organic extracts are filtered and then boiled for 12 hours under nitrogen. The chloroform extracts are stripped of solvent *in vacuo* to afford 0.3 g of a yellowish-brown oil, which may be triturated with the saturated $NaClO_4$. The solid that forms may be recrystallized from the ethanol-diethyl ether filtrate if an increased yield of 0.1 g is desired.

Treatment of the aqueous layer with 50 mL of a saturated sodium perchlorate solution affords a white precipitate that is collected and recrystallized from absolute ethanol-diethyl ether (1:3). The absolute ethanol should be added dropwise to a boiling mixture of the solid in diethyl ether until dissolution is apparent. A hot vacuum filtration (sintered-glass funnel; aspirator) should also be performed. The filtrate is then boiled and diethyl ether is added dropwise until cloudiness persists. Upon standing 24 hours, 7.1 g (90%) of 1-ethyl-1-phenylphospholanium perchlorate (mp 81-83°) precipitates. *Anal.* Calcd. for $C_{12}H_{18}ClO_4P$: P, 10.58. Found: P, 10.78.

Properties

1-Ethyl-1-phenylphospholanium perchlorate is a white solid that is slightly hygroscopic and slightly soluble in $CHCl_3$. Thus after recrystallization the sample should be dried over phosphorus pentoxide at 65° (0.2 torr) for 48 hours before analysis. Infrared absorption maxima (KBr wafer) occur at 3.42, 6.28 (vw), 6.95, 7.10, 9.13 (vs), 11.6, 12.64, 13.14, 13.35, and 14.37 μ, The proton NMR spectrum ($CDCl_3$, TMS internal standard) shows the following resonances: δ1.23 (doublet of triplets, J_{PCCH} = 20 Hz, J_{HCCH} = 8 Hz, CH_3CH_2P, 3H), 2.23 (multiplet, β-CH_2CH_2, 4H), 2.71 (multiplet $-CH_2PCH_2-$, PCH_2CH_3, 6H), 7.78 (multiplet, Ar$-H$, 5H). The ^{31}P NMR spectrum (40.5 MHz) showed a ^{31}P resonance at δ−51.4 (12% in $CDCl_3$) relative to 85% H_3PO_4.

References

1. K. L. Marsi, *Chem. Commun.*, **1968**, 846; *J. Am. Chem. Soc.*, **91**, 4724 (1969); K. L. Marsi and R. T. Clark, *J. Am. Chem. Soc.*, **92**, 3791 (1970); W. Egan, G. Chauviere, K. Mislow, R. T. Clark, and K. L. Marsi, *Chem. Commun.*, **1970**, 733.
2. P. Gillespie, F. Ramirez, I. Ugi, and D. Margquarding, *Angew. Chem. Int. Ed. Engl.*, **12**, 91 (1973) and references cited therein.
3. P. Beck, "Quaternary Phosphonium Compounds," in *Organic Phosphorus Compounds*, G. M. Kosolapoff and L. Maier, (eds.), Vol. 2, John Wiley-Interscience, New York, 1972, Chap. 4.
4. For a summary of the synthesis of phosphines see L. Maier, "Primary, Secondary, and Tertiary Phosphines," in *Organic Phosphorus Chemistry*, Vol. 1, G. M. Kosolapoff and L. Maier (eds.), John Wiley-Interscience, New York, 1972, Chap. 1; D. M. Hellwege and K. D. Berlin, in *Topics of Phosphorus Chemistry*, Vol. 6, Interscience Publishers, New York, 1969, pp. 1-186; T. E. Snider, C. H. Chen, and K. D. Berlin, *Phosphorus*, **1**, 81 (1971).
5. W. B. McCormack, *Org. Synth.*, Coll. Vol. 5, 787 (1973).
6. H. R. Hays and D. J. Peterson, "Tertiary Phosphine Oxides," in *Organic Phosphorus Compounds*, G. M. Kosolapoff and L. Maier (eds.), Vol. 3, John Wiley-Interscience, New York, 1972, Chap. 6.
7. K. B. Mallion and F. G. Mann, *J. Chem. Soc.*, **1964**, 6121.
8. K. Issleib and H.-R. Roloff, *Chem. Ber.*, **98**, 2091 (1965); A. Y. Garner and A. A. Tedeschi, *J. Am. Chem. Soc.*, **84**, 4734 (1962); F. G. Mann and M. J. Pragnell, *J. Chem. Soc.*, **1965**, 4120; K. B. Mallion and F. G. Mann, *Chem. Ind. (London)*, **1963**, 654; R. E. Goldsberry, D. E. Lewis, and K. Cohn, *J. Organometal. Chem.*, **15**, 491 (1968).

9. M. H. Beeby and F. G. Mann, *J. Chem. Soc.*, **1951**, 411; G. Märkl and K. H. Heier, *Angew. Chem. Int. Ed. Engl.*, **11**, 1016 (1972); G. Märkl, *Angew Chem. Int. Ed. Engl.*, **2**, 153 (1963); F. G. Holliman and F. G. Mann, *J. Chem. Soc.*, **1947**, 1634.
10. J. Meisenheimer, J. Casper, M. Höring, W. Lauter, L. Lichtenstadt, and W. Samuel, *Ann. Chem.*, **449**, 213 (1926).
11. W. R. Purdum and K. D. Berlin, *J. Org. Chem.*, **40**, 2801 (1975).
12. A. M. Aguiar, J. Beisler, and A. Miller, *J. Org. Chem.*, **27**, 1001 (1962).
13. R. S. Monson, "Advanced Organic Synthesis–Methods and Techniques", Academic Press, Inc., New York, 1971, p. 171.

34. BROMO FLUORO CYCLOTRIPHOSPHAZENES (Nongeminal Tribromotrifluoro and Dibromotetrafluoro Cyclic Triphosphonitriles)

Submitted by PHILIP CLARE* and D. BRYAN SOWERBY*
Checked by HARRY R. ALLCOCK† and ROBERT A. ARCUS†

Mixed halocyclotriphosphazenes with geminal structures can be prepared by fluorination of $P_3N_3Cl_6$ or $P_3N_3Br_6$ with potassium fluorosulfite, but the alternative, nongeminal compounds must be obtained by indirect methods. The starting materials are chloro dimethylamino cyclotriphosphazenes[1] with nongeminal structures, which can be obtained readily from hexachlorocyclotriphosphazene. The compounds are converted to the corresponding amino fluoro derivatives,[2,3] which in turn react with anhydrous hydrogen bromide to give the bromo fluoro derivatives.[4] Antimony trifluoride is used in the fluorination of $P_3N_3Cl_3[(N(CH_3)_2]_3$, but this reagent substitutes rapidly only the chlorine atoms in the $PCl[N(CH_3)_2]$ groups of $P_3N_3Cl_4[N(CH_3)_2]_2$ to give $P_3N_3Cl_2F_2[N(CH_3)_2]_2$. However, the two remaining chlorine atoms can be substituted using potassium fluorosulfite.

All experiments should be carried out in dry apparatus with dried solvents. With the exception of the mixed-halide final products, the compounds can be handled in the atmosphere, but the exposure time should be kept to a minimum.

A. 2,4,6-TRICHLORO-2,4,6-TRIS(DIMETHYLAMINO)-2,2,4,4,6,6-HEXAHYDRO-1,3,5,2,4,6-TRIAZATRIPHOSPHORINE AND 2,2,4,6-TETRACHLORO-4,6-BIS(DIMETHYLAMINO)-2,2,4,4,6,6-HEXAHYDRO-1,3,5,2,4,6-TRIAZATRIPHOSPHORINE

$$P_3N_3Cl_6 + 6(CH_3)_2NH \longrightarrow P_3N_3Cl_3[N(CH_3)_2]_3 + 3(CH_3)_2NH \cdot HCl$$

*Department of Chemistry, University of Nottingham, Nottingham, England.
†Department of Chemistry, Pennsylvania State University, University Park, PA 16802.

$$P_3N_3Cl_6 + 4(CH_3)_2NH \longrightarrow P_3N_3Cl_4[N(CH_3)_2]_2 + 2(CH_3)_2NH{\cdot}HCl$$

Procedure

A solution of 17.2 g (0.38 mole) of anhydrous dimethylamine in dry diethyl ether (50 mL) is added from a pressure-equalizing dropping funnel over 30 minutes to a well-stirred solution of 22.2 g (0.064 mole) of hexachlorocyclotriphosphazene in 750 mL of diethyl ether contained in a 1-L, three-necked flask cooled to -78°. The solution is then allowed to warm slowly overnight to room temperature when the precipitated amine hydrochloride is filtered and washed with diethyl ether. After removal of the solvent from the combined filtrate and washings, the material remaining is recrystallized from light petroleum ether (68-80° fraction). On careful fractionation the pure *trans*-isomer (mp 105°) can be obtained as the first fraction (13.6 g), but subsequent fractions are mixtures of the *cis*- and *trans*-isomers. Such mixtures, however, are acceptable for the fluorination described in Section B.

The procedure for the tetrachloro derivative is similar, except that 27.4 g (0.61 mole) of dimethylamine in 75 mL of diethyl ether is added to a solution of 53.1 g (0.15 mole) of the hexachloro derivative in 750 mL of diethyl ether at -78°. The product (35.4 g), after recrystallization from light petroleum ether (60-80° fraction), is a mixture of *cis*- and *trans*-isomers suitable for fluorination.

B. 2,4,6-TRIS(DIMETHYLAMINO)-2,4,6-TRIFLUORO-2,2,4,4,6,6-HEXAHYDRO-1,3,5,2,4,6-TRIAZATRIPHOSPHORINE AND 2,2-DICHLORO-4,6-BIS(DIMETHYLAMINO)-4,6-DIFLUORO-2,2,4,4,6,6-HEXAHYDRO-1,3,5,2,4,6-TRIAZATRIPHOSPHORINE

$$P_3N_3Cl_3[N(CH_3)_2]_3 + SbF_3 \longrightarrow P_3N_3F_3[N(CH_3)_2]_3 + SbCl_3$$

$$3P_3N_3Cl_4[N(CH_3)_2]_2 + 2SbF_3 \longrightarrow 3P_3N_3Cl_2F_2[N(CH_3)_2]_2 + 2SbCl_3$$

Procedure

■ **Caution.** *All reactions should be carried out in an efficient fume hood as chlorinated ehtanes are toxic, and the toxicity of the fluoro cyclotriphosphazenes has not been investigated.*

For the first compound a solution of 13.3 g (0.036 mole) of 2,4,6-trichloro-2,4,6-tris(dimethylamino)-2,2,4,4,6,6-hexahydro-1,3,5,2,4,6-triazatriphosphorine in 150 mL of 1,2-dichloroethane, previously dried for 24 hours over $CaSO_4$, is stirred efficiently with a Teflon-bladed paddle stirrer in a 250-mL, three-necked flask. The flask also carries a water-cooled condenser topped by a $CaCl_2$ drying tube. The solution is brought to reflux using an oil bath when 7.6 g (0.043 mole) of finely powdered SbF_3* is added rapidly.

The course of the reaction may be followed by gas-liquid chromatography using a silicone oil stationary phase. The reaction is complete after about 3 hours and the mixture is then filtered and the solvent removed under reduced pressure. An opaque gum remains and is extracted with five 100-mL portions of hot petroleum ether (40-60° fraction). About 1 mL of water is then added to the combined extracts and, on shaking, any antimony compounds remaining are precipitated. Filtration gives a clear solution to which is added another few drops of water to ensure completeness of separation of antimony residues. The water is removed and the solution is dried over anhydrous $CaCl_2$ for 2 hours. Removal of the petroleum ether under reduced pressure gives a colorless liquid of sufficient purity for the reaction with HBr (Sec. D).

An additional small amount of the product may be obtained by boiling the antimony residues with a solution of triethylamine (20 mL) in 200 mL of 40-60° petroleum ether for about 1 hour. The product is then isolated as described above, giving a total yield of 9.2 g (78%).

The apparatus and procedure used to obtain the difluoro derivative are very similar to those described above. A solution of 37.7 g (0.103 mole) of the tetrachlorobis(dimethylamino) derivative in 200 mL of 1,1,2,2-tetrachloroethane (dried over $CaSO_4$) is brought to reflux temperature and 14.0 g (0.078 mole) of powdered SbF_3 is added. Reaction is complete after about 6 hours of reflux and the product, $P_3N_3Cl_2F_2[N(CH_3)_2]_2$, is isolated as a colorless liquid from the mixture as described above; the yield is 29.5 g (86%).

Properties

Both the amino chloro starting materials and these amino fluoro derivatives are mixtures of *cis*- and *trans*-isomers. Fluorination of $P_3N_3Cl_3[N(CH_3)_2]_3$, however, gives a product containing about 95% of the *trans*-isomer and the reaction product may crystallize spontaneously after removal of the petroleum ether. If this does not occur, the pure isomer can be obtained as a white solid (mp 42°) by recyrstallization from hot 40-60° petroleum ether. The corresponding *cis*-isomer is less volatile and can be isolated from the mother liquors by preparative glc as a white solid melting at 35-36°. The mixture of *cis*- and *trans*-isomers of $P_3N_3Cl_2F_2[N(CH_3)_2]_2$ can be distilled at 73-5°/0.04 torr, but it is not possible to effect a separation of the isomers in this way. Separation is possible by pre-

*The reactivity of SbF_3 is reduced greatly by surface hydrolysis, but shorter reaction times with increased yields result when purified material is used. The crude trifluoro derivative is dissolved in anhydrous methanol and, after filtration to remove insoluble substances, the solvent is removed, first using a rotary evaporator and finally in a vacuum of 10^{-1} torr. The solid is then powdered in a dry box and kept in a desiccator.

parative glc which gives the liquid *trans*-isomer as the first fraction; the less-volatile *cis*-isomer is a solid (mp 54-55°).

C. 2,4-BIS(DIMETHYLAMINO)-2,4,6,6-TETRAFLUORO-2,2,4,4,6,6-HEXAHYDRO-1,3,5,2,4,6-TRIAZATRIPHOSPHORINE

$$P_3N_3Cl_2F_2[N(CH_3)_2]_2 + 2KSO_2F \longrightarrow P_3N_3F_4[N(CH_3)_2]_2 + 2KCl + 2SO_2$$

Procedure

A solution of 29.5 g (0.089 mole) of 2,2-dichloro-4,6-bis(dimethylamino)-4,6-difluoro-2,2,4,4,6,6-hexahydro-1,3,5,2,4,6-triazatriphosphorine in 200 mL of nitromethane, dried over $CaSO_4$, is refluxed with vigorous stirring for 20 hours with 40 g of a KSO_2F/KF mixture.* The solution is then filtered and the solvent is evaporated under reduced pressure. The residue is extracted with five 60-mL portions of petroleum ether (40-60° fraction) and the tetrafluorotriphosphazene is obtained as a liquid by evaporation of the solvent. The yield is 20 g (75%).

Properties

The product is again a mixture of *cis*- and *trans*-isomers that can be separated into the pure liquid isomers by preparative-scale glc.

D. 2,4,6-TRIBROMO-2,4,6-TRIFLUORO-2,2,4,4,6,6-HEXAHYDRO-1,3,5,2,4,6-TRIAZATRIPHOSPHORINE AND 2,4-DIBROMO-2,4,6,6-TETRAFLUORO-2,2,4,4,6,6-HEXAHYDRO-1,3,5,2,4,6-TRIAZATRIPHOSPHORINE†

$$P_3N_3F_3[N(CH_3)_2]_3 + 6HBr \longrightarrow P_3N_3F_3Br_3 + 3(CH_3)_2NH{\cdot}HBr$$

*The mixture, which contains about 20 mole% KSO_2F, can be prepared by shaking finely divided, anhydrous KF with an excess of liquid SO_2 in a sealed pressure vessel for 5 days. Pure KSO_2F is available commercially from, for example, PCR Inc., P.O. Box 1466, Gainesville, FL, or through use of an alternative preparative route[5]; a 5% excess over the stoichiometric quantity is required if the pure material is used.

†Analogous chloro fluoro cyclotriphosphazenes[4] can be obtained in a similar fashion using HCl in place of HBr. More rigorous precautions are necessary because higher pressures may be generated. The reactants can be sealed in a Carius tube at −196°, which is then placed in a steel bomb. As the tube warms to room temperature the bomb should be pressurized with an inert gas to compensate for the pressure generated inside the tube. As reaction proceeds this compensating pressure should be released gradually.

$$P_3N_3F_3[N(CH_3)_2]_2 + 4HBr \longrightarrow P_3N_3F_4Br_2 + 2(CH_3)_2NH \cdot HBr$$

The reactions are carried out in a thick-walled glass Carius tube (dimensions about 2.5 cm od × 20 cm) fitted with a "Rotaflo" TF 6/24 valve* and a standard taper joint for attachment to a vacuum line. Four grams of either $P_3N_3F_3[N(CH_3)_2]_3$ or $P_3N_3F_4[N(CH_3)_2]_2$ is added to the Carius tube, which is attached to a vacuum line and degassed. The line is equipped with a bulb of known volume and 6.5 g of HBr is condensed onto the phosphazene at liquid nitrogen temperature. After tightening the Rotaflo valve, the tube is detached from the line. (■ **Caution.** *As a safety measure, because high pressure may be generated, the tube is surrounded immediately by a protective metal gauze or placed in a metal tube. Use the fume hood.*)

The tube is allowed to warm to room temperature. The mixture liquefies and the pressure decreases, probably because of the formation of HBr addition compounds. After liquefaction it is safe to heat the tube in an oil bath to 100°. As reaction proceeds a second liquid layer of dimethylamine hydrobromide forms and reaction is complete within 36 hours.† The tube is then cooled to -45° and attached to a vacuum line (10^{-2} torr) for 3 hours to remove excess HBr. Each of the bromo fluoro cyclotriphosphazenes can then be isolated by allowing the tube to warm to room temperature and distilling the products into a trap held at -196°. Alternatively, the products can be vacuum distilled using conventional apparatus, but recovery is more efficient by the former method. The yield of $P_3N_3Br_3F_3$ is 3.8 g (70%) and that of $P_3N_3Br_2F_4$ is 4.5 g (90%).

Properties

The tri- and dibromo derivatives are volatile liquids with boiling points of 205 and 151°, respectively, at atmospheric pressure. They are readily hydrolyzed by atmospheric moisture and should be kept under nitrogen. They can be analyzed by conventional microanalytical methods for nitrogen. *Anal.* Calcd. for $P_3N_3Br_3F_3$: N, 9.7. Found N, 9.4. Calcd. for $P_3N_3Br_2F_4$: N, 11.3. Found N, 10.8. Assessment of purity, however, is best carried out by glc analysis followed by mass spectrometry. A 9-foot glass column packed with 10% silicone gum rubber on Chromsorb W is useful in verifying the absence of impurities after distillation, but this packing does not reveal the presence of the *cis-* and *trans-*isomers that are present in both liquids. Use of a stationary phase composed of 3% silicone oil and 13% dinonyl phthalate, however, gives sufficient separation

*Supplied by J. A. Jobling, Laboratory Division, Stone, Staffordshire, England. Alternatively, a Fischer-Porter 795-005 threaded glass valve may be used.

†Incomplete reaction with HBr leads to the formation of less-volatile amino bromo fluoro cyclotriphosphazenes that remain with the dimethylamine hydrobromide during the final separation step. This is most likely the cause of low yields.

in each case for use as a preparative method. In both cases the *cis*-isomer has a lower retention time.

Separated isomers can be distinguished by ^{19}F NMR spectroscopy (see Table III) or infrared spectroscopy. Major bands in the latter occur at the following wave numbers, *cis*-$P_3N_3Br_3F_3$; 1235 (vs), 968 (m), 859 (m), 750 (m), 644 (s), 513 (s); *trans*-$P_3N_3Br_3F_3$: 1233 (vs), 946 (m), 858 (s), 743 (w), 636 (w), 548 (vs), 517 (s); *cis*-$P_3N_3Br_2F_4$: 1264 (vs), 1247 (vs), 997 (m), 952 (s), 867 (m), 846 (s), 738 (m), 636 (s), 518 (s), 499 (s); *trans*-$P_3N_3Br_2F_4$: 1262 (vs), 1242 (vs), 995 (m), 960 (m), 948 (s), 935 (m), 870 (vs), 840 (vs), 737 (m), 618 (m), 548 (vs), 480 (s). Of the four compounds only the *cis*-tribromo derivative is a solid (mp 76°).

TABLE III ^{19}F NMR Data

	δ^a	J_{PF}(Hz)
cis-$P_3N_3Br_3F_3$	−22.2	1045[b]
trans-$P_3N_3Br_3F_3$	−18.0(1)[c]	1010
	−20.1(2)	1020
cis-$P_3N_3Br_2F_4$	−21.0(2)	1030
	−69.3(1)	920
	−71.5(1)	950
trans-$P_3N_3Br_2F_4$	−19.0(2)	1020
	−70.4(2)	910

[a]In ppm from $CFCl_3$.
[b]Real $^1J_{PF}$.
[c]Relative intensities

References

1. R. Keat and R. A. Shaw, *J. Chem. Soc.*, **1965**, 2215.
2. B. Green and D. B. Sowerby, *J. Chem. Soc. A.*, **1970**, 987.
3. B. Green, D. B. Sowerby and P. Clare, *J. Chem. Soc. A.*, **1971**, 3487.
4. P. Clare, D. B. Sowerby and B. Green, *J. Chem. Soc. Dalton*, **1972**, 2374.
5. F. Seel, *Inorg. Synth.*, 9, 113 (1967).

Chapter Eight

SULFUR COMPOUNDS

35. SILVER(I) SULFAMATE

$$2AgNO_3 + 2KHCO_3 \longrightarrow Ag_2CO_3 + 2KNO_3 + H_2O + CO_2$$

$$Ag_2CO_3 + 2NH_2SO_2OH \longrightarrow 2NH_2SO_2OAg + H_2O + CO_2$$

Submitted by G. C. BRITTON*
Checked by DONALD HANKUS† and C. DAVID SCHMULBACH†

Silver(I) salts are normally characterized by their insolubility in water, the only common soluble compounds being the nitrate, chlorate and perchlorate.

Silver(I) sulfamate is remarkable for its considerable solubility in water over a wide range of temperatures and the fact that the anion is normally nonoxidizing. The substance is also a very important starting material for numerous syntheses of other sulfamates, imidodisulfates, and their derivatives.

The first recorded synthesis of silver(I) sulfamate was that by Berglund[1] in 1876; he used the very slow reaction between silver(I) sulfate and barium(II) sulfamate. Later methods have employed the reaction between sulfamic acid and silver(I) oxide[2] or silver(I) carbonate[3]; an electrolytic process has also been described.[4]

*Middleton St. George College of Education, Darlington, England.
†Department of Chemistry and Biochemistry, Southern Illinois University, Carbondale, IL 62901.

The synthesis offered below is probably best in terms of availability of starting materials and ease of their purification, using, as it does, silver(I) nitrate, potassium hydrogen carbonate, and high-purity commercial sulfamic acid.

Procedure

■ **Caution.** *Because of the photosensitivity of silver(I) salts, all processes described below must be carried out expeditiously in subdued light.*

Seventeen grams (0.1 mole) of silver(I) nitrate is dissolved in 100 mL of warm distilled water. To the solution is added 10 g (0.1 mole) of potassium hydrogen carbonate in 40 mL of warm distilled water, with attention being paid to the evolution of carbon dioxide gas.

The light-yellow precipitate of silver(I) carbonate that forms is washed twice with 100 mL of warm distilled water by decantation, and a solution of 10 g (0.1 mole) of sulfamic acid in 100 mL of distilled water at 60° is added carefully. After the reaction has subsided, the solution of silver(I) sulfamate is heated briefly to 100° to expel dissolved carbon dioxide and is then filtered rapidly into a crystallizing dish. The solution is allowed to stand overnight in the dark.

The next morning the crop of long (up to 3 cm), transparent needles that has deposited is filtered using sintered glass, washed with a little cold distilled water, and air-dried. If the substance is required completely water-free, further drying can be effected *in vacuo* over concentrated sulfuric acid or in an Abderhalden dryer at 100°. The filtrate is set aside to evaporate further, and if care is taken to avoid mechanical losses, the yield is usually 90% or slightly more.

Alternatively, when the precipitated silver(I) carbonate from the first stage of the preparation has been washed by decantation, a solution of 10 g (0.1 mole) of sulfamic acid in 50 mL of hot, distilled water is carefully added. The concentrated solution of silver(I) sulfamate is filtered rapidly with suction and, on cooling, deposits a large crop of tiny, transparent crystals. Purification is achieved by recrystallization from warm water. *Anal.* Calcd. for NH_2SO_2OAg: Ag, 52.89. Found: Ag, 52.73 (Volhard method).

Properties

Silver(I) sulfamate forms colorless, rhombic crystals that darken slowly on exposure to light. The substance has the following solubilities in water.[5]

T(deg)	0	25	30	40	50	60	100
S,g/100 mL H_2O	2.25	8.09	9.35	13.26	16.80	22.7	~78.0[6]

Solutions are stable to boiling for short periods. On heating, solid silver(I) sulfamate decomposes at 201° (rapidly at 230°), first forming disilver(I) imidodisulfate $(AgSO_2O)_2NH$, then silver(I) sulfate

$$3(AgSO_2O)_2NH \longrightarrow 3Ag_2SO_4 + 3SO_2 + NH_3 + N_2$$

A number of double salts are known[2]

The infrared spectrum shows absorption at the following frequencies (in cm^{-1}): 3307 (m), 3265 (m), 1600 (w), 1546 (m), 1217 (s), 1123 (m), 1053 (s), 954 (m), 802 (s), 595 (m), 585 (s), 561 (s), 498 (m).[6]

Acknowledgment

The author wishes to thank Dr. J. B. Wilford and Mr. J. Speakman of Teesside Polytechnic, Middlesbrough, Cleveland, England for their invaluable assistance with the recording of infrared spectra.

References

1. S. Berglund, *Acta Lund,* **13**, 77 (1876).
2. P. Sakellaridis, *Compt Rend.*, **247**, 2142 (1958).
3. L. Bicelli and A. La Vecchia, *Ann. Chim. (Rome)*, **46**, 351 (1956).
4. S. Tajima, Y. Hiratsuka, Y. Kosaka, and Y. Ishii, *J. Electrochem. Soc. Jap.*, **22**, 2111 (1954).
5. M. Odehnal, *Chem. Listy*, **49**, 1571.
6. G. C. Britton, Unpublished observations.

36. HEPTATHIAZOCINE(heptasulfurimide) and TETRABUTYLAMMONIUM TETRATHIONITRATE

Submitted by J. BOJES,* T. CHIVERS,* and I. DRUMMOND*
Checked by RALF STEUDEL† and FRITZ ROSE†

The well-established synthesis of cyclic sulfur imides from disulfur dichloride and ammonia in *N,N*-dimethylformamide[1] requires close attention and gives rise to S_7NH, $S_6(NH)_2$ isomers, and $S_5(NH)_3$ isomers in approximate yields of

*Department of Chemistry, The University of Calgary, Calgary T2N 1N4, Alberta, Canada.
†Institut für Anorganische und Analytische Chemie, Technischen Universität Berlin, Strasse des 17 Juni 135, 1 Berlin 12, Germany.

12-21, 7, and 0.2-0.4% (based on S), respectively. A more convenient procedure for the preparation of S_7NH alone in yields up to 41% (after chromatographic separation) is the reaction of sodium azide with elemental sulfur in hexamethylphosphoric triamide.[2] The deep-blue intermediate, which is dramatically apparent in the synthesis of cyclic sulfur imides, has been identified as the $[NS_4]^-$ ion.[3,4] This ion is isolated as its tetrabutylammonium salt from the reaction of S_7NH with $[(C_4H_9)_4N]OH$ in diethyl ether at low temperature.

A. HEPTATHIAZOCINE(Heptasulfurimide)

$$7S + NaN_3 + H_2O \longrightarrow S_7NH + N_2 + NaOH$$

Procedure

■ **Caution.** *Hexamethylphosphoric triamide is suspected to be carcinogenic. Inhalation or skin contact should be avoided.*

Hexamethylphosphoric triamide (HMPA, $[(CH_3)_2N]_3PO$; Aldrich; 500 mL) is dried with NaH and then sufficient K metal is added to turn the batch completely blue. The solvent is distilled (bp 68°/0.4 torr[5] or 115°/15 torr[6]) under nitrogen; the first 30 mL and the last 70 mL are rejected.* The product should be stored in a dry, oxygen-free atmosphere. Sodium azide (19.0 g, 0.292 mole) and sulfur (20.0 g, 0.623 mole) in HMPA (400 mL) are placed in a 1-liter flask, which is then purged with nitrogen. (It is necessary to work in an N_2 atmosphere, even though N_2 is evolved during the reaction, particularly for smaller-scale reactions.) The outlet of the flask is connected to a gas bubbler and the reaction mixture is stirred (Teflon-coated magnetic stirrer) at room temperature and rapidly becomes inky blue. The reaction is continued until evolution of nitrogen ceases (about 3 days). The reaction mixture is then added to an equal volume of ice-cold 10% hydrochloric acid. The buff-yellow precipitate is allowed to settle (1½ hr), filtered, washed with water, and dried in a vacuum desiccator over Drierite to give 20.2 g of crude product. The crude product should not be allowed to stand for more than 1 day, otherwise decomposition occurs. This product is ground finely in a mortar and extracted with eight successive 50-mL portions of anhydrous diethyl ether. The ether extracts are combined in a Petri dish and the ether is allowed to evaporate in the draught of a fume hood to give 11.3 g of yellow crystals containing sulfur imides mixed with sulfur. This

*When HMPA is degassed to remove dimethylamine[7] (N_2, 3 hr; then under vacuum, 2 hr), but not dried and redistilled, the yield of S_7NH after chromatography is 6.0 g, 32%. The checker reports that the distillation procedure is unnecessary provided that a previously unopened bottle of HMPA is used.

mixture is dissolved in the minimum amount of *redistilled*, reagent grade CS_2 and chromatographed on a silica gel column (60-200 mesh, 5 cm × 22 cm fitted with Teflon stopcock) using CS_2 as the liquid-phase eluant.* Fractions (about 20 mL) are collected in weighed 50-mL beakers using a drop rate of about 200 drops/min, and the eluate is tested for the presence of S_7NH by shaking with KOH in methanol. All the imides give a characteristic violet color. The fractions are left to evaporate in the fume hood to give 2.0 g of sulfur, 7.7 g of S_7NH, and 0.03 g of $S_6(NH)_2$ isomers. The purity of the chromatographed fractions can be checked using thin-layer chromatography, with 20 × 5 cm strips of Eastman Chromogram sheets. More complete details of the chromatographic separation of sulfur and sulfur imides are given in reference 1. The product is recrystallized from CCl_4 to give 6.0 g of very pale yellow crystals (mp 113°, lit. mp 133.5°[1]). *Anal.* Calcd. for S_7NH: N, 5.86. Found: N, 5.66. The yield of recrystallized material is 32%. The infrared spectrum[8,9] of CS_2 solutions shows bands at 3332 and 807 cm^{-1}

B. TETRABUTYLAMMONIUM TETRATHIONITRATE

$$S_7NH + [(C_4H_9)_4N]OH \longrightarrow [(C_4H_9)_4N][NS_4] + 3S + H_2O$$

Procedure

Heptathiazocine (1.4 g, 5.85 mmole) is dissolved in anhydrous diethyl ether (150 mL) in a 250-mL, three-necked flask. The solution is cooled to −78° (acetone-Dry Ice) under nitrogen and tetrabutylammonium hydroxide 7.5 mL of 25% solution in methanol, Eastman, 7.25 mmoles) in diethyl ether (12.5 mL) is added slowly. A yellow precipitate (sometimes with a greenish tinge) is formed immediately. The solution is kept at −78° and filtered through a coarse glass frit in a low-temperature filter funnel (Kontes, airless-ware).† The greenish-yellow solid on the frit is transferred quickly to a dry box, and in the course of about 1 day it becomes a deep-purple solid. This solid contains elemental sulfur (x-ray

*An alternative purification procedure that avoids the time-consuming chromatographic step is suggested by the checker. The dry, crude product is extracted with boiling methanol. The solution is cooled to 25° and sulfur is filtered quickly. Further cooling to −25° causes the crystallization of S_7NH, which is filtered and recrystallized to give S_7NH (containing a little S) suitable for preparative work. The overall yield is not changed when this purification procedure is used, but it is recommended that the chromatographic step be employed when pure S_7NH (free from S) is required.

†Rapid filtration through a coarse glass frit attached to one neck of the flask can be successful, but if the filtration is slow and the solution is allowed to warm a blue oil is formed.

powder pattern), which can be removed by Soxhlet extraction with dry *n*-pentane (300 mL) under nitrogen in a fritted-glass thimble. The extraction should be continued for 2 days to ensure complete removal of sulfur. Final purification is achieved by dissolving the dark-blue solid in tetrahydrofuran (THF) (10 mL), filtering, and allowing the THF to evaporate to give an oil. Addition of anhydrous diethyl ether (200 mL) precipitates $[(C_4H_9)_4N][NS_4]$ as a fine blue-black powder (mp 49°). The yield is 1.6 g, 71%. *Anal.* Calcd. for $C_{16}H_{36}N_2S_4$: C, 49.95; H, 9.43; N, 7.28; S, 33.34. Found: C, 50.20; H, 9.19; N, 7.12; S, 33.02. All manipulations of $[(C_4H_9)_4N][NS_4]$ are performed under a dry oxygen-free atmosphere.

Properties

The properties of S_7NH are well documented, and complete infrared and Raman spectral assignments have been published.[8,9] Tetrabutylammonium tetrathionitrate is moisture sensitive and is hydrolyzed by acids to give S, S_7NH, and $S_6(NH)_2$ isomers.[4] It can be stored for long periods of time in a dry oxygen-free atmosphere at room temperature. The salt is very soluble in THF but almost insoluble in diethyl ether. A 10^{-4} *M* solution in either THF or HMPA shows a characteristic visible absorption band at 595 nm (ϵ = 19,000).[4] The infrared spectrum (Nujol mull) shows bands at 1243, 1220, 1102, 1021, 922, 885, 737, 710, 608 (sh), 594, 560, 530 (sh), 510 (sh), 410 cm^{-1}.

References

1. H. G. Heal and J. Kane, *Inorg. Synth.*, **11**, 184 (1968).
2. J. Bojes and T. Chivers, *Inorg. Nucl. Chem. Lett.*, **10**, 735 (1974); *J. Chem. Soc., Dalton*, **1975**, 1715.
3. T. Chivers and I. Drummond, *J. Chem. Soc., Chem. Commun.*, **1973**, 734.
4. T. Chivers and I. Drummond, *Inorg. Chem.*, **13**, 1222 (1974).
5. T. Chivers and I. Drummond, *Inorg. Chem.*, **11**, 2525 (1972).
6. H. Normant, *Angew. Chem. Int. Ed.*, **6**, 1046 (1967).
7. T. Chivers and I. Drummond, *J. Chem. Soc., Dalton*, **1974**, 631.
8. J. Nelson, *Spectrochim. Acta*, **27A**, 1105 (1971).
9. R. Steudel, *J. Phys. Chem.*, **81**, 343 (1977).

INDEX OF CONTRIBUTORS

SUBJECT INDEX

Names used in this Subject Index for Volume XVI, as well as in the text, are based for the most part upon the "Definitive Rules for Nomenclature of Inorganic Chemistry," 1957 Report of the Commission on the Nomenclature of Inorganic Chemistry or the International Union of Pure and Applied Chemistry, Butterworths Scientific Publications, London, 1959; American version, *J. Am. Chem. Soc.*, 82, 5523-5544 (1960); and the latest revisions [Second Edition (1970) of the Definitive Rules for Nomenclature of Inorganic Chemistry]; also on the Tentative Rules of Organic Chemistry–Section D; and "The Nomenclature of Boron Compounds" [Committee on Inorganic Nomenclature, Division of Inorganic Chemistry, American Chemical Society, published in *Inorganic Chemistry,* 7, 1945 (1968) as tentative rules following approval by the Council of the ACS]. All of these rules have been approved by the ACS Committee on Nomenclature. Conformity with approved organic usage is also one of the aims of the nomenclature used here.

In line, to some extent, with *Chemical Abstracts* practice, more or less inverted forms are used for many entries, with the substituents or ligands given in alphabetical order (even though they may not be in the text); for example, derivatives of arsine, phosphine, silane, germane, and the like; organic compounds; metal alkyls, aryls, 1,3-diketone and other derivatives and relatively simple specific coordination complexes: *Iron, cyclopentadienyl-* (also as *Ferrocene*); *Cobalt(II), bis*(2,4-*pentanedionato*)- [instead of *Cobalt (II) acetylacetonate*]. In this way, or by the use of formulas, many entries beginning with numerical prefixes are avoided; thus *Vanadate (III), tetrachloro-*. Numerical and some other prefixes are also avoided by restricting entries to group headings where possible: *Sulfur imides*, with formulas; *Molybdenum carbonyl,* $Mo(CO)_6$; both *Perxenate,* $HXeO_6^{3-}$, and *Xenate (VIII),* $HXeO_6^{3-}$. In cases where the cation (or anion) is of little or no significance in comparison with the emphasis given to the anion (or cation), one ion has been omitted; e.g., also with less well-known complex anions (or cations): $CsB_{10}H_{12}CH$ is entered only as *Carbaundecaborate*(1–), *tridecahydro-* (and as $B_{10}CH_{13}-$ in the Formula Index).

Under general headings such as *Cobalt(III) complexes* and *Ammines*, used for grouping coordination complexes of similar types having names considered unsuitable for individual headings, formulas or names of specific compounds are not usually given. Hence it is imperative to consult the Formula Index for entries for specific complexes.

As in *Chemical Abstracts* indexes, headings that are phrases are alphabetized straight through, letter by letter, not word by word, whereas inverted headings are alphabetized first as far as the comma and then by the inverted part of the name. Stock Roman numerals and Ewens-Bassett Arabic numbers with charges are ignored in alphabetizing unless two or more names are otherwise the same. Footnotes are indicated by *n.* following the page number.

FORMULA INDEX

The Formula Index, as well as the Subject Index, is a cumulative index for Volumes XVI, XVII, and XVIII. The chief aim of this index, like that of other formula indexes, is to help in locating specific compounds or ions, or even groups of compounds, that might not be easily found in the Subject Index, or in the case of many coordination complexes are to be found only as general entries in the Subject Index. *All* specific compounds, or in some cases ions, with definite formulas (or even a few less definite) are entered in this index or noted under a related compound, whether entered specifically in the Subject Index or not.

Wherever it seemed best, formulas have been entered in their usual form (i.e., as used in the test) for easy recognition: Si_2H_6, XeO_3, NOBr. However, for the less simple compounds, including coordination complexes, the significant or central atom has been placed first in the formula in order to throw together as many related compounds as possible. This procedure often involves placing the cation last as being of relatively minor interest (e.g., alkali and alkaline earth metals), or dropping it altogether: MnO_4Ba: $Mo(CN)_8K_4 \cdot 2H_2O$; $Co(C_5H_7O_2)_3Na$;$B_{12}H_{12}O$. Where there may be almost equal interest in two or more parts of a formula, two or more entries have been made: Fe_2O_4Ni and $NiFe_2O_4$; $NH(SO_2F)^{2-}$ $(SO_2F)_2NH$, and $(FSO_2)_2NH$ (halogens other than fluorine are entered only under the other elements or groups in most cases); $(B_{10}CH_{11})_2Ni^{2-}$ and $Ni(B_{10}CH_{11})^{2-}$.

Formulas for organic compounds are structural or semistructural so far as feasible: $CH_3COCH(NHCH_3)CH_3$. Consideration has been given to probable interest for inorganic chemists, i.e., any element other than carbon, hydrogen, or oxygen in an organic molecule is given priority in the formula if only one entry is made, or equal rating if more than one entry: only $Co(C_5H_7O_2)_2$, but $AsO(+)\text{-}C_4H_4O_6Na$ and $(+)\text{-}C_4H_4O_6AsONa$. Names are given only where the formula for an organic compound, ligand, or radical may not be self-evident, but not for frequently occurring relatively simple ones like C_5H_5 (cyclopentadienyl), $C_5H_7O_2$ (2,4-pentanedionato), C_6H_{11} (cyclohexyl), C_5H_5N (pyridine). A few abbreviations for lignads used in the test, including macrocyclic ligands, are retained here for simplicity and are alphabetized as such, "bipy" for bipyridine, "en" for ethylenediamine or 1,2-ethanediamine, "diphos" for ethylenebis(diphenylphosphine) or 1.2-bis (diphenylphosphino)ethane or 1.2-ethanediylbis(diphenylphosphine), and "tmeda" for N,N,N',N'-tetramethylethylenediamine or N,N,N',N'-tetramethyl-1,2-ethanediamine.

Footnotes are indicated by *n*, following the page number.